EVOLUTION BY NATURAL SELECTION

Confidence, Evidence and the Gap

Species and Systematics

The *Species and Systematics* series will investigate the theory and practice of systematics and taxonomy and explore their importance to biology in a series of comprehensive volumes aimed at students and researchers in biology and in the history and philosophy of biology. The book series will examine the role of the study of biological diversity at all levels of organization and focus on the philosophical and theoretical underpinnings of research in biodiversity dynamics. The philosophical consequences of classification, integrative taxonomy and future implications of rapidly expanding data and technologies will be among the themes explored by this series. Approaches to topics in *Species and Systematics* may include detailed studies of systematic methods, empirical studies of exemplar taxonomic groups, and historical treatises on central concepts in systematics.

For more information visit:
www.crcpress.com/browse/sense/crcspeandsys

Evolution by Natural Selection

Confidence, Evidence and the Gap

MICHAELIS MICHAEL

CRC Press is an imprint of the
Taylor & Francis Group, an **informa** business

CRC Press
Taylor & Francis Group
6000 Broken Sound Parkway NW, Suite 300
Boca Raton, FL 33487-2742

© 2016 by Taylor & Francis Group, LLC
CRC Press is an imprint of Taylor & Francis Group, an Informa business

No claim to original U.S. Government works

International Standard Book Number-13: 978-1-4987-0087-0 (Hardback)

Visit the Taylor & Francis Web site at
http://www.taylorandfrancis.com

and the CRC Press Web site at
http://www.crcpress.com

I dedicate this book to the memory of my father, Stephanos Michael (July 25, 1927–January 23, 2015). I wanted to finish this book in time to show him; sadly, he died before I could finish it. My father was a strange and wonderful man. He grew up on a farm in Cyprus. From the age of twelve, he was taken out of school and was not allowed to realize his own aspirations to get an education. He raised a pair of oxen from calves and trained them to the plow. Barefoot, he plowed the family farm. He travelled across the world to Australia, a place he'd barely heard of. He made sure that I got the education he missed out on. I grew up in a major city, but many weekends were spent in the bush with him. As the song says, "As a boy he'd take me walking by mountain field and stream, and he'd show me things not known to kings, and secret between him and me." I miss him.

Contents

Series Preface

The *Species and Systematics* series is a broad-ranging venue for authors to provide the scientific community with comprehensive treatments of the theoretical underpinnings, history, and philosophy of fundamental concepts in systematic biology and the science of taxonomy. The series also intends to link the historical landscape of ideas to new technology and expanding information in order to stimulate discussion among students and researchers in biology about the future course we are charting in biodiversity research.

There are many approaches to the study of biological diversity, and to embrace this, future volumes in *Species and Systematics* may include detailed development and comparisons of existing and novel methods in systematics and biogeography, empirical studies of exemplar organisms that provide new insight into old questions and raise new questions for biologists and philosophers of science, and historical treatises on central and reoccurring concepts that benefit from both a retrospective and a new perspective. Some volumes will address a single important concept at great depth, giving authors the freedom to present ideas with their own slant, while others will be edited collections of shorter papers intended to place alternative views in sharp contrast.

One of the most important tasks for philosophers and historians is to *stress test* our terms and concepts so that we are pushed to reevaluate what we know or think we know. To draw on popular culture, the film character Inigo Montoya's now meme-famous saying sums this up well, "You keep using that word, I do not think it means what you think it means" (*The Princess Bride*, 1987). This volume contributed by Michaelis Michael is a call to all biologists to consider if they really know what they mean when using or discussing one of our most central ideas—Darwin's theory of natural selection. From different vantage points, Michael shows how easy it is to use natural selection and, simultaneously, how hard it can be to directly access particulars. Readers are guided through carefully laid-out arguments that test the mettle of natural selection and put a spotlight on strengths and weakness, encouraging precise usage. In this way, *Evolution by Natural Selection: Confidence, Evidence and the Gap* is a valuable contribution that hits the mark intended for the *Species and Systematics* series.

Kipling Will
Berkeley, CA
September 9, 2015

Acknowledgments

The first and last person to thank is my wife, Aislinn Batstone. She has been my rock. I love and appreciate her, and this would not have happened without her. My kids, Stefan, Druce, and Nora, have kept me on the straight and narrow and reminded me what life is all about. I would also like to thank Malte Ebach for encouraging me to write this book and for being a consistent sounding board and Chuck Crumly and Jennifer Ahringer at Taylor and Francis for being so helpful and accommodating. Colleagues and friends have been very good to me, in particular, Karyn Lai, Damian Grace, Steve Cohen, Debra Aarons, Mengistu Amberber, Matthew Mison, Andrew Horne, Jim Franklin, Markos Valaris, and Melissa Merritt. I have been very lucky to have valued and talented friends in James Bucknell, Adam Dickerson, Amitavo Islam, and Max Rabie, who have been very helpful in providing comments and encouragement. Andrew Westcombe stepped in and helped enormously with the final proof reading; his help was generous and beyond the call of friendship. In 2001, I was invited to visit the Department of Philosophy at Wake Forest University as a Thomas Jack Lynch Visiting Associate Professor, where I presented some of this material. I thank the department and especially Prof. Ralph Kennedy, who has become a very good friend and collaborator. I have been sitting on a lot of this material for a very long time. Tom Rich had encouraged me to publish it many years ago, and so did Allen Hazen. Both were transplanted Americans living in my home town of Melbourne, Australia, and both gave me good advice that I took far too long to follow.

Author

 **Michaelis Michael** studied Zoology and Philosophy at Monash University in Melbourne, Australia, before achieving a PhD in Philosophy at Princeton University. He works across a number of areas in philosophy, from human rights to formal logic. He has recently published articles and contributed chapters on the role of noncognitive factors in religious conversion, on the metaphysics of mind, and against the idea that we need to adopt deviant logics to deal with inconsistent theories in science.

1 Introduction

Today it might seem that evolutionary theory, particularly in its Darwinian form, is inescapable. The geneticist Dobzhansky (1973) famously said that nothing in biology makes sense except in the light of evolution. In biology, it is certainly everywhere. In fact, the theory has broken out of its biological stable and is now being used to shed light on areas that have had little contact with biological thinking. The theory of natural selection is deployed to explain everything from changing our understanding of illness to shedding light on political allegiances, the nature of voting patterns of fiction and arts (Tooby and Cosmides, 2001), Shakespeare's plays (Orians, 2014), and the tendency toward sexual violence (Liddle et al., 2012). This comes at the same time as a revolution in technology, which allows us to map the genotype of individuals relatively easily. The genomic revolution goes hand in hand with the theory of natural selection. Added to that is a bewildering ocean of data threatening to swamp us. No longer can we eyeball data and get an idea of what it is indicating. Now we must have recourse to data mining programs that look for correlations in the masses of data gathered by many hands. These three factors together have changed biology. This book is mostly on the Darwinian theory and its proper understanding. It cannot help but advert to the genetic revolution. The progress we have made in understanding the genetics of development and protein synthesis is truly amazing. The amount of information we are gathering simply cannot be comprehended. It is beyond anyone's ability to encompass. If we look at the ubiquity of the theory, then it would look as though there was not much to say about it really.

This book is about Darwin's theory of natural selection: what the theory says, what it is aimed at explaining, and how it manages to explain.* The problem faced in writing it is that it can seem as though there was no need to write it: Darwin's theory has a familiarity that means that everyone believes that they already understand it perfectly. Darwin's theory has the ring of the obviousness about it, even if that became apparent only with hindsight. There are many who agree with Jacques Monod's quip that the curious thing about evolutionary theory is that everyone does think they understand it perfectly (Monod, 1974). The fact that there are so many divergent views about the theory means that not everyone who thinks they understand the theory perfectly can be right. As David Hull (1995) put it, "Evolutionary theory seems so easy almost anyone can misunderstand it." Hull's note of caution seems right, and this is even in a context in which evolutionary theory is so familiar that grade school children are introduced to it and the theory is rapidly being deployed to explain many and diverse aspects of the world. Why be cautious? The reason is there are serious questions about the theory and its deployment that we need to get clear about. The theory aims to give us historical–causal explanations. It is aimed as explaining not just how things are but why they are that way. This is part

---

* Darwin's writings are helpfully very readily available online in searchable format at http://darwin -online.org.uk/contents.html.

of its theoretical appeal. Hence, how does the theory manage to explain, and what do we need to know before we can use it to explain?

As anyone who has used the theory of natural selection knows only too well, there are two salient aspects of this theory when used to explain phenomena. The first is we use it with ease; we can come up with explanations using the theory of natural selection with almost embarrassing ease. In fact, this ease got lampooned by Stephen Jay Gould and Richard Lewontin (1979) as just-so stories. It can seem that taking any feature of the world we care to mention, we can always come up with a story that picks out some aspect of that feature that makes it look plausible wherein the aspect confers adaptive significance and explains why that feature is there. The second that follows readily from the first is that if there are all these different explanations available, how are we going to be able to tell if we have found the right one? That is, there is a real problem trying to justify any use of the theory of natural selection. Thus, on the one hand, using the theory is easy; on the other hand, using it is really hard. These two aspects are two sides of the same coin, and any attempt to make the theory perspicuous has to explain both aspects. Also, it must explain the reason why it is so hard to establish the correctness of any use of the theory about our justification in deploying the theory to explain the world.

With this in mind, what I will do is try to show that there really are issues we do not understand very clearly and that these issues lead to problems in our uses of the theory. To do that, I will show just how an old argument against the theory of natural selection can be used to show us both how the theory explains what it does and what the theory needs by way of conceptual resources to achieve that. I am going to suggest that the history of evolutionary theory has been a sequence of theoretical developments, not all of which can be thought of as an improvement. In particular, I want to suggest that by taking our eye off the main game of how we are explaining using the theory of natural selection, we reshape the concepts to make them easier to apply to the world but end up stripping the theory of natural selection of its explanatory power. We end up with a theory that spins free of the sort of empirical friction required for good science. Curiously, a close reading of Darwin's own presentation of the theory shows how he was sensitive to the manner in which the theory had to be phrased to allow it to play that explanatory role.

The biological world seems to have plenty of functions. Aristotle describes these functions as what something is for 'the sake of', that is, have a purpose or goal. He argued these functions provided an argument against the theory of natural selection.

A difficulty presents itself: why should not nature work, not for the sake of something, nor because it is better so, but just as the sky rains, not in order to make the corn grow, but of necessity? What is drawn up must cool, and what has been cooled must become water and descend, the result of this being that the corn grows. Similarly if a man's crop is spoiled on the threshing-floor, the rain did not fall for the sake of this—in order that the crop might be spoiled—but that result just followed. Why then should it not be the same with the parts in nature, e.g. that our teeth should come up of necessity—the front teeth sharp, fitted for tearing, the molars broad and useful for grinding down the food—since they did not arise for this end, but it was merely a coincident result; and so with all other parts in which we suppose that there is purpose? Wherever then all the parts came about just what they would have been if they had come be for an end, such things survived, being organized spontaneously in a fitting way; whereas those which grew otherwise perished and continue to perish, as Empedocles says his 'man-faced ox-progeny' did. (Aristotle, *Physics*, Book 2, Chapter 8)

Aristotle, who was no slouch as a biologist, thought it was plain that incisors and molars had different functions; incisors "fitted for tearing and molars broad and useful for grinding down food." Then almost surreally, Aristotle considers a theory about the origin of the differently shaped teeth. The theory he considers is a theory we would be tempted to call, and that Darwin himself recognized to be, the theory of natural selection.* What if, Aristotle considered, organisms with teeth that did cut and grind efficiently in this way survived and passed on these features? Aristotle argued that that theory could not explain the advent of functionality, the fact that differently shaped teeth have different functions. Right or not, this extraordinary event in the history of thought will give us some clues about what the theory of natural selection seeks to explain and how it might do that. This book aims to answer these questions.

This might seem obvious. After all, the theory of natural selection is so obvious that Thomas Huxley famously reflected,

> I suppose that Columbus' companions said much the same when he made the egg stand on end. The facts of variability, of struggle for existence, of adaptation to conditions, were notorious enough; but none of us had suspected that the road to the heart of the species problem lay through them, until Darwin and Wallace dispelled the darkness, and the beacon-fire of the 'Origin' guided the benighted. (Huxley, 1901, 1: 183)

Even schoolchildren (perhaps not enough of them) learn about the theory of natural selection, and many high-quality documentaries have also had a really important role in spreading knowledge of the theory of natural selection. No doubt many contemporary practicing biologists were first inspired as children by David Attenborough's wonderful series *Life on Earth*, which ran on BBC Two from January to April 1979.

There is an interesting battle for the heartland of evolutionary theory taking place. On one side is the flexible might of the modern information revolution with its computer models and its data mining techniques. On the other side is the detailed biological work developed by functional morphologists. This is the background of the intellectual battle between the population genetics influenced Richard Dawkins and the paleontologist Stephen Jay Gould. In the most salient book on the debate, Kim Sterelny's (2001) *Dawkins vs. Gould*, there is sadly not much discussion of the backgrounds from which Gould and Dawkins come. The debate between Gould and Dawkins marks a fundamental theoretical disagreement.

Sterelny describes this as a debate about levels of selection and optimality. Certainly, optimality plays a role, but it is more fundamentally about the ways we explain evolutionary biology and what levels causally explain the sorts of change over time in the biota that Darwin's theory tried to explain. Gould was a pluralist about levels of selection, and Dawkins is famous for advocating the gene-centric conception of evolution. Sterelny himself is much closer to the viewpoint articulated by Dawkins.

---

* Darwin himself identifies Aristotle as a precursor to his work on natural selection in the historical introduction to *The Origin*.

Gould was an invertebrate paleontologist. His PhD was on West Indian land snails and their evolution. As a part of that research, he developed his theory of punctuated equilibria, the idea that Darwin was wrong in thinking that evolution was gradual and constant. Gould suggested that the evidence he had gathered suggested that in fact evolution varied in time and tempo. With Niles Eldredge, he published a very important paper in 1972 suggesting that species will show little change over the majority of their history, a period of stasis (Eldredge and Gould, 1972). When things do change, they happen very abruptly, from a geological time scale instantaneously. This attack on Darwin's gradualism was seen as revolutionary at the time. Now it is fair to say that it has been absorbed and seen to be in fact consistent with the work of the doyen of evolutionary thought, Ernst Mayr (1954), who developed a theory of allopatric speciation decades earlier. His work was also aligned in particular with the work of Sewall Wright, a seminal population geneticist who predicted both stasis and a significant role for allopatric speciation, in particular, in peripheral isolated populations where evolution would be quicker and mechanisms such as genetic drift would be strongest (Wright, 1942).

Of course, the benefit of hindsight makes Gould's and Eldredge's ideas seem less radical and fit easily into the intellectual landscape. However, it is certainly the case that they ruffled theoretical feathers with their papers and swam against the intellectual tide. Eventually, the tide turned, but at the time, they did not garner much support. From the present perspective, it is hard to see what the fuss was about.

Gould was also significant in evolutionary theory for a paper he wrote with Richard Lewontin (1979), "The Spandrels of San Marco and the Panglossian Paradigm: A Critique of the Adaptationist Programme." This paper sought to show that the idea that the world was optimal was not an empirical truth discoverable in inquiry but rather a presupposition that was impossible to test. Moreover, Lewontin and Gould argued that there are reasons to doubt that it makes sense to regard the outcomes of evolution by natural selection as optimal. A third way in which Gould was important in evolutionary theory is his work on bottlenecks, on catastrophic events that lead to mass extinctions. Gould's greatest impact was probably in the presentation of science to nonscientists. In this respect, he is similar to Dawkins. Gould was a wonderful stylist, and his essays for the *Natural History* magazine were collected in a series of books beginning with *Ever Since Darwin* until his death, which were in the U.S. bestsellers and significantly raised the profile of evolutionary science among the public. Gould's public profile may have misled the public into thinking that his ideas were received as wisdom among contemporary evolutionary scientists. This really was not so.

The overwhelming consensus view about evolutionary science was derived not from the paleontologist Gould but from the American population geneticist G. C. Williams and the British mathematically inclined theoretical biologist W. D. Hamilton. Williams and Hamilton represented the mainstream, and Dawkins learned from them the gene-focused conception of natural selection. Therefore, Dawkins and not Gould represented orthodox evolutionary thinking. Dawkin's first important book, *The Selfish Gene* (Dawkins, 1976), was an explicit popularization and extension of Hamilton's ideas.

What are the lines of conflict in this war, and how do these battles play out? One aspect of it is the way we make discoveries in evolutionary science. Evolutionary science is our attempt to chart the history of the biological world and to provide explanations that give as understanding not just of what happened when but why that happened. That contrast between what happened and why it happened is important. We are not merely charting the history as a sequence of events (difficult as that is), but we are attempting the more important task of explaining why that sequence took place.

Later we will see how these issues of optimality and levels of causal explanation are connected. Explaining the natural world is the aim of biological science, and it is now a cliché that evolution is at the basis of all explanations of the biological world. Moreover, although clichés only become clichés because there is an important kernel of truth to them, just what that kernel might be is not clear.

Charles Darwin made his theory public in 1858. As a gentleman member of the Royal Society, he had already been taken to the bosom by the scientific establishment and was famous among scientists and laypeople for his writings, particularly for his account of his worldwide voyage aboard the bark *HM Beagle*. He had quietly developed his theory of evolution by natural selection from the late 1830s after that expedition. He wrote huge amounts; the posthumously published version of what became Chapter 4 of *The Origin* ran longer than *The Origin* itself. He did not make his theory known to his confidantes for some time but rather honed his presentation of the theory over a period of twenty years before bringing it into the public realm. Then came the letter from a little known animal collector suffering from a tropical fever in the jungles of the Malay Archipelago. The letter was famously a shock. The writer, Alfred Russel Wallace, was a poor but hardworking naturalist who supported himself by his writings and by collecting biological specimens. He had come up with a theory that explained how organisms could change to new conditions. Like Darwin, Wallace had read by curious coincidence Thomas Malthus's 1798 book *An Essay on the Principle of Population*, which suggested that population growth put relentless pressure on resources, giving rise to a competition to survive. Inspired by Wallace's letter, Darwin hurried to get his theory into the public realm. When he did so, it was a rush job; a paper was presented alongside Wallace's paper to a meeting of the Linnean Society in 1858 and it raised little response. In fact, the Presidential Report in 1858 presented in May 1859 indicated, "The year which has passed has not, indeed, been marked by any of those striking discoveries which at once revolutionize, so to speak, the department of science on which they bear." Knowing what was to happen, we would beg to differ. Actually, it did not take too much hindsight: November 1859 saw the publication of the first edition of *The Origin of Species* (Darwin, 1859), and the reactions were anything but muted. It sold out straight away, and the reactions of biologists were overwhelmingly, but not exclusively, positive. However, there were persistent qualms over the structure of the theory of natural selection. Is the theory just circular? Is it just pseudo-science? These worries were related to the empirical character of the theory, and they were not without some foundation. In the following chapters, I shall try to show how these misgivings can be responded to. The misgivings can be assuaged, but in doing so, we must understand the ready tendencies which cause these objections.

As part of showing what these tendencies are, I aim to show just why it is at once so easy and so hard to do evolutionary biology. It is easy, on the one hand, because evolutionary explanations for phenomena are seemingly endless in their variety and number, making them readily available to those with enough imagination. On the other hand, doing evolutionary biology is ridiculously hard because it is so hard to establish any particular explanation for a phenomenon. This feature of the theory of evolution by natural selection needs explaining. One explanation of this feature, an explanation that threatens the legitimacy of the theory, is that the theory is just bogus. The theory would then not really be providing explanations at all. I shall show why this complaint is not idle as the theory of evolution by natural selection is sometimes presented. However, I shall argue that Darwin's theory in its modern articulation is one of the fundamental steps in our understanding of the world. It can be presented in a manner that shows the objections to its conceptual foundations to be misplaced.

Chapter 2, *The Circularity Argument*, presents the circularity objection, showing how it undermines the theory and has to be taken seriously. Chapter 3, *Resolving the Problem of Circularity*, shows how to articulate the theory of natural selection so that it both explains what Darwin was trying to explain and also escapes from the problematic circularity objections. This leads to an attempt to outline the conceptual foundations of the theory of evolution by natural selection, but it will not address issues of the empirical implementation of the theory. These issues, once highlighted and separated from the conceptual issues, will be the focus of Chapters 6 and 7. Once we see how the theory is conceptually constructed, it will become more or less obvious why the ease and difficulty we noted are the features we should expect for this vitally important theory.

WHAT IS EVOLUTION?

We now think about the world as changing, whether at the level of the universe as a whole—the cosmological changes from the Big Bang—or at the human level—as we develop different ways of living and technology that changes our world. There was a time when people thought of the world as unchanging and eternal. Before we launch into trying to understand Darwin's distinctive theory, we should pause briefly to consider a phenomenon that his theory sought to explain—evolution itself.

What concerns us is evolution, simply put, the change of the biological world over time. That is very inclusive and probably too inclusive. Many biological changes are just not of interest to us. The change from one individual to another that looks exactly alike, such as what we might find in a population of bacteria, does involve change but not the sort of change we are interested in. We are interested in the way organisms change in character. Again the way a caterpillar changes into a moth does involve a change in character over time but within the life of one individual. That too is not evolution of the sort that interests us. Of course, that is something we have discovered and is not so by definition. Evolution is quite consistent with the inheritance of acquired characteristics. It is after all the way heritable characteristics change over time that is of central interest in the study of evolutionary biology. If organisms did inherit acquired changes or some of them, evolution would have to incorporate such acquired changes. Some recent advocates of inheritance of acquired characteristics have alleged that this is contrary to Darwin's theory. More properly, this is contrary to the neo-Darwinian synthesis but specifically to the Mendelian aspects of that synthesis. Many people seem to have forgotten that Darwin himself thought that there were cases of inheritance of acquired characteristics. In Chapter 7, *Heritability of Characteristics: The Role of Genetics and Epigenetics*, we will see just what the theory of Natural Selection needs by way of heredity. The real contrast between Darwin and Lamarck was whether there is a goal for evolution and whether that goal provided any explanation of the changes of organisms toward the goal. Lamarck held that there was such a goal explaining changes along the road to it. Darwin held there was no such goal.

When we consider the changes that the biological world undergoes, which changes are we to focus on, any changes or all changes? Is the mere change of location of the same organisms over time an example of evolution? The characteristics of organisms change during their lives; organisms grow to maturity, for example. Is it change in phenotypic characteristics in general? Is it perhaps not just any change in phenotypic characteristics but just those that are heritable? Is it not tied to change in phenotypic characteristics at all but rather the genetic makeup of the biota which is really at issue? Surprisingly enough, I do not think we need to make up our minds here and now on the important changes that are of interest to us. We should note that all these changes can and do occur, and they are liable to various sorts of explanation. What is of interest to us is which of these changes are liable to explanation by Darwin's theory of natural selection, and further, there is the important question of the extent to which the theory of natural selection actually does provide the correct explanation for the phenomena we observe. If we use the most inclusive notion of evolution as change in the biota over time, then we can see just how much of the change in the biota over time is able to be explained by the theory of natural selection.

WHAT WAS DARWIN TRYING TO EXPLAIN?

The theory of evolution is so fundamental to our contemporary understanding of the world that it is hard but worthwhile to try to ask what it was that Darwin was trying to explain. This requires the mental effort to see something as curious and worthy of reflection that we take to be obvious. It might be surprising to see that Darwin developed the theory of evolution by natural selection to explain two phenomena that are not themselves evolutionary phenomena. However, once we see what the theory of natural selection was trying to explain, we can start to understand how it was supposed to do that and see whether it succeeded or not. Darwin's theory is, to my mind, one of the greatest scientific achievements of his or any age. It changed our sense of ourselves and provided insights as deep as any we can get. Yet it has not been a smooth ride for the theory. From within and without science, it has met with objections and incredulity. This book is a kind of justification and an explanation for that theory. The aim is to lay out the theory of natural selection in a way that allows us to see how it can explain what it is trying to explain. We will see that there are numerous ways of spelling out the theory and not all of them are happy. Some will lead to theories that do not and cannot possibly provide the explanations we want

the theory to provide. Thus, why define key terms in those ways? We will see just
how the temptations to define a thing in those mistaken ways are motivated by some
otherwise unimpeachable scientific motivations. Losing sight of the big picture leads
to changes in definitions, which scupper our hopes of an explanatory theory.

Darwin sought to explain two phenomena that were as plain as the nose on his
face and which other thinkers had overwhelmingly disregarded. The two phenomena
seem obvious to us today and would have seemed so obvious in his day too. It took
a rare feat of genius to see that they were in need of an explanation: diversity and
adaptation.

DIVERSITY

Organisms come in many varieties. Darwin's question here is How is it that there
are so many kinds of organisms? The world might have been populated with just one
kind of organism but it is not. Training as he was for the clergy at Cambridge, Darwin
learned at the university that the Bible taught that each type of creature was specially
and separately created by God. The variety of organisms is bewildering. They come
in all sorts of shapes, sizes, and modes of life, and they differ in their characteristics.
This variation in the characteristics among the kinds of organism leads to the second
obvious phenomenon Darwin saw needed an explanation: although organisms vary
in their characteristics, organisms are adapted to their mode of life.

ADAPTATION

If the first phenomenon, the diversity among organisms, was obvious, the second
phenomenon, adaptation, is just as visible. Organisms are well suited to their way
of life. What could be more obvious than that? Fish that live in the open ocean have
different characteristics from fish that live in estuaries or swamps. Each has characteristics that suit it to its way of life. It does not need to be said that a fish is better
adapted to life in water than is a dog. This is one of the key phenomena Darwin
sought to explain.

Darwin was not the first to notice that organisms were adapted to their environment. While at Cambridge studying for the clergy, he had become captivated by the
works of someone who had made much of just that fact, Reverend William Paley.
Paley is today famous for the development of what is called "natural theology," the
doctrine that finds arguments for the existence of an omnipotent benevolent creator
in the details of the natural world. Paley pointed to the many and varied ways in
which organisms have characteristics that help them fit into their mode of life. Were
we to chance on a watch, he said, we would marvel at the intricate ways the parts fit
together so as to allow it to tell the time. We could not help but see the hand of an
intelligent creator shaping the parts and their relations. Examining the even more
complex biological machines and the way their parts hang together to allow them
not merely to survive but to flourish in so many different ways once again, we find
ourselves drawn to an even greater designer. This argument was not an argument
essentially about biology. It was in fact an argument from the presence of adaptation
to the existence of a divine creator. Moreover, it is important to see that Darwin was

entranced when he studied this. He could see the adaptation organisms displayed, and the minutest study seemed to find more and more adaptations. It is indeed striking that adaptations do not give out as we find out more about the world. It can be said that hemoglobin is an adaptation for transporting oxygen and carbon dioxide and that DNA transcription depends on particular enzymes that play specific roles in that transcription, roles without which life itself would be hard to imagine. In other words, the more we study organisms, the more adaptations we find.

However, long before the good Reverend Paley developed his argument, a philosopher had argued that the biological world showed signs of the purposive. Aristotle, the great Greek polymath, argued that the natural world was full of purpose (Aristotle, *Physics*, Book II, Chapter VIII). As we saw earlier in his discussion of how incisors are sharp for the sake of cutting food and molars are flat for the sake of grinding it. Teeth are differently shaped because of their purposes. Thus, Aristotle also saw that this idea of purpose and the way it fitted organisms to their mode of living was an important fact about the world. In fact, it seems Aristotle thought that the theory which explained how things were by a series of causes and effects, with the best fitting ones surviving and reproducing their kind, cannot account for the purpose we perceive in the world and so must be a mistaken account of the biological world.

The important thing to notice is that this notion of purposiveness, that is, adaptation, is understandable and noted in the world even in the absence of an explanation of it. Paley and Aristotle each discerned it no less than Darwin himself. Indeed, even someone who thought that the entire world began a year ago as a curious quantum singularity can see that, surprisingly enough, organisms are well suited to their way of life. For that reason, this fit between an organism's characteristics and its way of life cannot be conceptually tied to the theory of natural selection. We do not need to have conceived of that theory, let alone believe it, to understand that organisms are well adapted to their way of life. This crucial fact shows us that we cannot define adaptation in terms natural selection.

So with these phenomena—diversity and adaptation—crying out for an explanation, eventually Darwin thought he had the answer: the theory of natural selection.

This gives us several reference points that will help us to make sense of the theory of natural selection; the theory Darwin sought to develop had to be the sort of theory that could conceivably play a role in explaining these two phenomena. When we say this, we are not requiring that the theory of natural selection is the unique explanation for the diversity and adaptation of organisms, but rather, it has to be the sort of theory that could play that role, just as God's beneficence and omnipotence could have played that role.

REFERENCES

Darwin, C. 1859. *On the Origin of Species*. 1st ed. London: John Murray; 2nd ed., 1860; 3rd ed., 1861; 4th ed., 1866; 5th ed., 1869; 6th ed., 1872; 6th ed., with additions and corrections, 1876.
Dawkins, R. 1976. *The Selfish Gene*. Oxford: Oxford University Press.
Dobzhansky, T. 1973. Nothing in biology makes any sense except in the light of evolution. *American Biology Teacher* 35: 125–129.

Eldredge, N. and S. J. Gould. 1972. Punctuated equilibria: An alternative to phyletic gradualism. In *Models in Paleobiology*, edited by T. J. M. Schopf. San Francisco: Freeman Cooper. pp. 82–115. Reprinted in N. Eldredge, *Time Frames*. Princeton: Princeton University Press, 1985, pp. 193–223.

Gould, S. J. and R. C. Lewontin. 1979. The spandrels of San Marco And the Panglossian paradigm: A critique of the adaptationist programme. *Proceedings of the Royal Society of London B* 205: 581–598.

Hull, D. 1995. Universal Darwinism. *Nature* 377: 494.

Huxley, L., ed. 1901. *The Life and Letters of Thomas Henry Huxley*. Vol. 1. London: Macmillan.

Liddle, J. R., T. K. Shackelford, and V. A. Weekes-Shackelford. 2012. Why can't we all just get along? Evolutionary perspectives on violence, homicide, and war. *Review of General Psychology* 16(1): 24–36.

Malthus, T. 1798. *An Essay on the Principle of Population*. London: Joseph Johnson.

Mayr, E. 1954. Change of genetic environment and evolution. In *Evolution as a Process*, edited by J. Huxley, A. C. Hardy, and E. B. Ford. London: Allen and Unwin. pp. 157–180.

Monod, J. 1974. *On the Molecular Theory of Evolution*. Reprinted in *Evolution*, edited by M. Ridley. New York: Oxford University Press, 1997, p. 389.

Orians, G. H. 2014. *Snakes, Sunrises, and Shakespeare: How Evolution Shapes Our Loves and Fears*. Chicago: University of Chicago Press.

Sterelny, K. 2001. *Dawkins vs. Gould*. Cambridge.

Tooby, J. and L. Cosmides. 2001. Does beauty build adapted minds? Towards an evolutionary theory of aesthetics, fiction and the arts. *SubStance* 94/95: 6–27.

Wright, S. 1942. Statistical genetics and evolution. *Bulletin of the American Mathematical Society* 48: 223–246.

2 The Circularity Argument

I returned, and saw under the sun, that the race is not to the swift, nor the battle to the strong, neither yet bread to the wise, nor yet riches to men of understanding, nor yet favour to men of skill; but time and chance happeneth to them all. (Ecclesiastes 9:11)

Charles Darwin first publicly aired his new theory of evolution by natural selection in a paper read to the Linnean Society in 1858 (Darwin and Wallace, 1858). Darwin himself was not there and neither was the author of the other paper on the same topic, Alfred Russel Wallace. The papers were read and ignored. The President of the Linnean Society delivered a report that characterized 1858 as a year in which little of significance had occurred in the realm of science. Whether he had misread the mood of the scientific community, or whether the wave of interest in Darwin's ideas had not yet developed, what is clear is that fifteen months later, the scientific community was indeed ready for Darwin's ideas. *The Origin of Species*, published on November 24, 1859, sold out on that very first day. However, even after he published it, he did not see *The Origin* as his definitive presentation of his theory. Throughout his life, he worked on what he called his "big book." This was supposed to be the book of which *The Origin* was a mere abstract. He never published his big book. Over a century later, some parts of the manuscript intended to be that book did get published and what we see is that the abstract is a clearer presentation of the theory than his big book would have been. What was concise, sharp, and focused in *The Origin* was obscured in the big book, hidden as it was in the wealth of detailed evidence he had brought together over the decades. *The Origin*'s crucial fourth chapter, called "Natural Selection," which is less than forty pages in most editions, has its correlate in the big book that, when published, turned out to be the same size as the whole of *The Origin* itself.

Natural selection was Darwin's big insight. It was the idea that led to Huxley's famous remark "How stupid not to have thought of that!" Natural selection is an idea so simple, so powerful, so obvious that like many brilliant breakthroughs, thinking our way to the view of those who do not grasp it is as difficult as any act of imagination. There are so many of these that the claim that the theory of evolution by natural selection is either a tautology, or that it involves a circularity, has been made repeatedly since Darwin first proposed his theory. Some, perhaps most, have raised it as an objection seeing it as exposing the emptiness of Darwin's theory and a telling refutation of Darwin's theory, and others like Waddington, or the late nineteenth century philosopher C.S. Peirce, have regarded it positively as Darwin's true theoretical innovation. The argument which is supposed to establish the claim is simple and at first glance seems convincing. Unfortunately, we do not know Darwin's own reaction to the argument or his reaction to the claim. However, the claim has been rejected by many and resurrected by many. It's the zombie argument allegedly residing at the heart of evolutionary theory. However, sensible folks know that zombies don't exist, so it is time the argument is laid to rest. The argument, if it were correct, far from being a mere restatement of the theory of natural selection, would leave the theory without explanatory power. Once we see that the argument can be laid to rest, there is still a very pressing question: if the argument is so clearly wrong-headed, why do so many definitions of the key concepts of the theory of natural selection lead to this objection? The explanation for its zombie-like character is partly due to the fact that the theory of natural selection is not well understood, and partly due to the fact that the theory of natural selection is not well explained by scientists. It may seem audacious for a philosopher to make claims about the understanding scientists have of one of their basic theories, but the arguments set out below will justify these claims.

I think the circularity argument can be successfully rebutted and is actually a very useful prop that can help us understand the theory of natural selection by showing how the elements of the theory combine to provide its explanatory structure. Moreover, far from being merely of historical interest, the circularity objection does indeed have a purchase on some current explanations of the theory of natural selection. This objection is a threat to various putative explications of the theory by both philosophers and biologists. The circularity objection will not turn out to be a real threat to the theory as it should be understood nor indeed as the theory tends to be used in practice by biologists, but it lies in wait for those less cautious with their concepts.

THE SIMPLE CIRCULARITY ARGUMENT

1. The fittest are, by definition, those which survive.
2. Natural selection is, by definition, the survival of the fittest.
    Therefore, by substitution of definitions.
3. Natural selection is the survival of those which survive.

This argument is in the terminology familiar to Darwin in the mid-nineteenth century and sounds a little archaic today. This is the simplest form of the argument but, even here, there are some contemporary biologists who fall foul of it.

Is it right that the fittest are the survivors? Just what does survival entail? And just what is it that is the fittest? Are we talking about individuals, characteristics, genes, alleles, or species? Darwin couched his theory in terms of individuals. So for him and for many, until the genetic perspective arose, the theory of natural selection is essentially about the properties of individuals and their survival to reproduce.* Darwin distinguished *natural selection*, which was sensitive to the fitness of individuals from *sexual selection*, which was sensitive to the attraction of individuals to others. There are two ways to think of this: The first is as a sequence of filters—natural selection first filtering the range of variation and then sexual selection operating on

---

* However, as we will see below, the theory does not actually depend on the theory being couched in terms of individuals. The argument above is presented in a way that is neutral about these matters. It does not really matter whether we are talking about individuals or genes but once we are clear what we do mean, what we mean by "survival" will be different.

the resultant restriction of the variation. The second is as separate forces directing change in a population, in which these forces can be in conflict. Later, we will look at this more closely but here we can ignore the complications introduced by separating natural selection from sexual selection.

If fitness was about a relationship to the environment, attraction had little to do with that and could run counter to it. It is interesting to see that Darwin was thinking of two quite different processes: the one explaining how individuals manage to survive to the age of reproduction; the other picking among those reproductively available individuals. In this way, the features that help individuals survive can make that individual less attractive to breed with, or conversely, characteristics that are harmful when it comes to surviving can come into their own as attractions. The peacock's gaudy and impractical tail makes him less fit but more sexually attractive to peahens in Darwin's way of thinking. Modern biological thought rolls both of these types of selection into one type of selection. It is important to see the difference between the two sorts of selection to get a sense of how Darwin was thinking of things.

MODIFIED VERSIONS OF THE CIRCULARITY ARGUMENT

The terms in which the simple circularity argument is conducted can seem anachronistic and irrelevant to contemporary biological thinking. The language is intentionally tied to Darwin's own. Indeed, even in Darwin's day, the idea that survival was an all-or-nothing affair was not plausible. The lessons to be gleaned by the simple circularity argument are nevertheless important and stand as correct even if we complicated the situation.

One way the simple version can seem false to reality is that a situation in which all the survivors are the fit variations and that the less fit just don't survive is, after all, rare. Some of the survivors are not the fittest and some of the fittest are not survivors. Today, a more common characterization of the fittest is that they are the ones that differentially survive; that is, disproportionately more of the fit survive as compared to the less fit. This gives us a new version of the circularity argument:

1. The fittest are, by definition, those which differentially survive.
2. Natural selection is, by definition, the differential survival of the fittest.
   Hence,
3. Natural selection is the differential survival of those which differentially survive.

This version of the argument is not merely of historical interest. It is quite common, for example, in population genetics to hear both that natural selection is the differential survival of alleles and that the fitness is a measure of differential reproductive success, so that the fittest alleles are those which differentially survive. Here then we have statements offered as definitions of the notions of natural selection and fitness, respectively. This really does lead to the conclusion that natural selection is just the differential survival of those alleles that differentially survive.

Contemporary biologists are more likely to recognize their concepts in a version of the argument phrased in terms of reproductive success:

1. The fittest are, by definition, those with the greatest reproductive success.
2. Natural selection is the differential reproductive success of the fittest.
3. Natural selection is the differential reproductive success of those with the greatest reproductive success.

Once again, we seem to reach a disturbing conclusion. Is that all there is to this much lauded theory?

We might think that this is misguided; that surely, there is more to being the fittest than merely surviving, even merely surviving differentially. One obvious suggestion is that being fittest is tied to degree of adaptedness. After all, it is the way organisms fit into their environment that explains their survival.

Now we can construct the following argument:

1. The fittest are, by definition, those that are best adapted.
2. Natural selection is the survival of the fittest.
   Thus,
3. Natural selection is the survival of the best adapted.

So far, so good; but the question immediately arises: what does it mean to be best adapted? One mistake to make here, but it is a tempting one, is to identify the best adapted as those which differentially survive. The temptation is easy to feel: by identifying degree of adaptation with rates of survival, we turn degrees of adaptation into a quantity we can measure. Measure the rates of survival and you have a measure of the degree of adaptation. However, defining the notion of adaptation like that leads to trouble.

1. The fittest are, by definition, those that are best adapted.
   The best adapted are those that differentially survive.
2. Natural selection is the differential survival of the fittest.
   Hence,
3. Natural selection is differential survival of those that differentially survive.

Once again, at first blush, this looks like it should be a worry.

There are many things in common in these various arguments: they depend on the definition of the key theoretical concepts of natural selection and fitness. First, we will see why some have thought that this is indeed all there is to the theory, whether they go on to say that it is a problem for the theory or not.

BAD REASONS TO DISMISS THE CIRCULARITY ARGUMENT

Are these various arguments really so worrying? Some scientists and others have thought not. One reason to think not would be if the arguments were no more than restatements of Darwin's thinking and, as such, really no objection to Darwin's theory at all. The philosopher C.S. Peirce is one.

> The Darwinian controversy is, in large part, a question of logic. Mr. Darwin proposed to apply the statistical method to biology... In like manner (to statistical reductions of gas thermodynamics) Darwin, while unable to say what the operation of variation and natural selection in any individual case will be, demonstrates that in the long run they will, or would adapt animals to their circumstances. (Peirce 1877, p. 7 in Buchler 1955)

Another is the geneticist C.H. Waddington.

> Darwin's major contribution was, of course, the suggestion that evolution can be explained by the natural selection of random variations. Natural selection, which was at first considered as though it were a hypothesis that was in need of experimental or observational confirmation, turns out on closer inspection to be a tautology, a statement of an inevitable although unrecognized relation. It states that the fittest individuals in a population (defined as those which leave most offspring) will leave the most offspring. Once the statement is made, its truth is apparent. This fact in no way reduces the magnitude of Darwin's achievement; only after it was clearly formulated, could biologists realize the enormous power of the principle as a weapon of explanation. (Waddington, 1960)

Peirce was no biologist but was an eminent mathematician and logician whose views on the logic of Darwin's theory might be less reliable. Waddington, on the other hand, was an important evolutionary biologist who made long-lasting contributions to the field. His work on canalization and the importance of development is becoming more and more important in modern biology. If Waddington can be so sanguine about this issue, is it really so bad if the theory of natural selection really were to fit this model?

The tendency not to worry much about the circularity argument is shared by some eminent philosophers of biology as well, and some philosophers fall afoul of the argument in their endeavor to explain away the apparent teleology in evolutionary biology.* Eminent philosopher of biology, Elliot Sober, is another who does not think these arguments raise serious worries about Darwin's theory. He makes two claims. First, that the conclusion derivable from the theory is not a logical tautology and, second, even if the conclusion were a tautology, having a tautology derivable from your theory is not a problem because, as a matter of logic, tautologies are derivable from every theory. Let's look at these two claims.

Logical tautologies are empty and implied by any theory, so Sober is correct that every theory will imply tautologies.† It is not so much that a tautology is derivable from the theory that is the problem. In fact, as Sober indicates, logical tautologies are derivable from all theories. He is absolutely right about that and the fact that this is not the problem at all. Moreover, Sober is correct that the claim that "the survivors survived" is not a tautology. It is not a matter of logic that there were any survivors

---

* For example, Ruth Millikan and Karen Neander.
† There are deviant logicians who dispute this. We need not dispute it here and, in any case, I suspect it is the deviants who are wrong on this score.

at all, so the claim that the survivors survived has to be a substantive claim because it implies that there were survivors and that is a substantive claim.

Elliot Sober suggested that the circularity argument should not worry evolutionary biologists. He points out that because the conclusion implies that something survives, it is not a tautology. Moreover, he argues that even if it were a tautology, that a theory entails a tautology or analytic statement is no criticism of the theory (Sober, 1993). The two claims are true. The conclusion is not, strictly speaking, a tautology, and every theory—true, false, significant, or insignificant—entails tautologies. However, is he right that these points mean that the circularity argument can be set aside? I want to explain why he is wrong about that.

Sober is correct that the worry does sometimes get put in terms of "the fittest differentially survive" being a logical tautology (Peters, 1976). Tautologies are true in all possible situations. In particular, a tautology cannot entail anything that might be false. The conclusion of the circularity argument

c. The survivors survived.

is indeed not a logical tautology because that sentence implies that there are survivors and there might not have been. This shows that (c) is not a logical truth. There is a recherché issue here of whether (c) should be understood as entailing that there are some survivors that we can pass over. Logical orthodoxy has it that it does not but the issue can be conceded to Sober because not too much is riding on it. We will see that his response does not get to the heart of the matter.

Every sentence entails a logical truth and any tautology is a logical truth. Therefore, any sentence entails any logical truth. Were the conclusion a tautology, then it would be utterly uninteresting that any particular theory entails it because every sentence entails it. Were he correct about the force of the objection from circularity, then Sober would have disarmed it. In the next section, we will see that this is not the real objection in the worry about circularity.

THE REAL PROBLEM RAISED BY THE ARGUMENTS: NO EXPLANATIONS

The real worry raised by this objection is not that the conclusion is a logical tautology. For neither (a) nor (b) is a logical tautology: we can imagine situations in which each is false. Rather, it is that the conclusion would amount to showing that evolution, or differential survival, and evolution by natural selection were equivalent. The worry raised is not that nothing is said in saying that x evolved through natural selection, for such a claim entails the substantial claim that x did indeed evolve. The real worry is that no more is added by the phrase "by natural selection." Although it is not a logical tautology, this does make the conclusion of the argument fall under what is meant by a tautology in common usage (Burchfield, 2000). The purported explanation of an episode of evolution by adverting to natural selection is empty. Sober seems to miss this altogether, but it is a serious worry.

Even though Elliot Sober seems to think that his response dissolves the problem, his observation does not get at the heart of the problems raised by the circularity objection. The key problem is that

 a. The hairy variety evolved.
 b. The hairy variety evolved by natural selection.

are importantly different claims. Moreover, the second claim is the stronger claim that entails the first claim but not vice versa. It does not follow as a simple matter of logic that if the hairy variety evolved, that it evolved by natural selection. We are offering an explanation in the second of how the hairy variety evolved. Yet if the two definitions offered in "The Simple Circularity Argument (page 10)" were accepted, the claim that the hairy variety evolved by natural selection would follow by logic alone from the claim that it evolved. And that really is unacceptable.

If the hairy variety differentially survived then, by the first definition, they were the fittest. However, because the fittest differentially survived according to the second definition, then that is an example of natural selection. Contrary to our first reaction, this shows that (a) implies (b). Now this is enough to show that insofar as we thought that by offering (b), we are providing an explanation of how (a) came about, we are mistaken. We say no more in (b) than we did in saying (a) and this, everyone will concede, is absurd.

The key problem raised is this: If the definitions (1) and (2) were accepted, then the qualifying phrase "by natural selection" would not contribute any meaning to (b): the two sentences, (a) and (b), would be logically equivalent—neither would say more about the world than the other. However, we do think that when we say that the hairy variety evolved by natural selection, we are saying more than just that the hairy variety evolved. The real problem with accepting (1) and (2) is that there would be simply no additional information in (b) than in (a). In particular, although we might have thought that (b) was explaining how it came to be that the hairy ones came to predominate, it would turn out that (b) merely repeats the claim that the hairy ones came to predominate, in which case (b) offers no explanation at all.

This is why the circularity argument cannot be embraced as spelling out Darwin's views, contrary to Peirce and Waddington, and, contrary to Sober, cannot simply be ignored. The circularity argument does indeed show that certain common ways of defining key concepts in the theory threaten to empty the theory of its explanatory power. Were the definitions accepted, the theory would not offer explanations at all but, as we saw, the whole point of the theory of natural selection is to provide an explanation, a mechanism that can explain the sorts of change associated with the two phenomena Darwin tried to explain: the variation and degree of adaptation manifested by the biological world. Because being able to provide such an explanation is the whole point of the theory, we must say that some of the definitions used in the argument must be rejected. To define both fitness and natural selection in the manner used in the simple circularity argument, or its more modern analogue, is to strip the theory of natural selection of any explanatory power.

This problem is not just an artifact of the way the theory of natural selection was presented. There are real issues with the ways that key terms are defined in evolutionary theory.* There are reasons for this that we will deal with in due course. However, the problems should have signaled that even very important concepts such as genetic drift would become absolutely inapplicable in a conceptual space dominated by concepts defined and used as they are in these arguments.

Genetic drift is the result of a sampling effect introduced into the genetic history of a population by the formation of gametes in sexually reproducing populations. Because each gamete will only have half the chromosomes of the parent and, in a successful production of an offspring, that offspring will have half of each parent's, it is possible that the "sample" of the parents' genes in the successful gametes is different in frequency of alleles than the parents' genetic frequency. Because the gamete must misrepresent the diploid frequency of an allele for which that individual is heterozygous, either by containing (thereby misrepresenting it as though it were homozygous for it) or by omitting it (thereby misrepresenting it as if it lacked it all together).

Genetic drift can lead to changes in the genetic constitution of the population. Yet if an allele were to become more common due to this chance sampling, that allele would have differentially survived. There has been a change and that allele's frequency has increased, that is, it differentially survived. From this observation, it follows that if the allele differentially survives, it is being selected. However, in that case, there is no room for a nonselective mechanism of evolution such as genetic drift under this suite of definitions. And that is a problem.

The key use of the theory of natural selection is in explaining how things came to be the way they are. However, if the premises of the circularity argument were correct, the theory would be utterly unfit for that purpose. It would provide no explanations of that sort at all.

When presented with this sort of argument, scientists tend to display skepticism. Even scientists who offer the equivalent of (1) and (2) as definitions are not immune to this response. This seems to be an underpinned conviction that there is no problem at all in continuing to use the theory in the way they were taught to use it. In part, this sort of reaction is the typical skeptical reaction from the nonphilosopher to philosophical quandaries. It is not obvious that this is a tendency to deplore. Indeed, philosophical quandaries, although often very simply stated, require subtlety and great sophistication to resolve. Witness, for example, the ease with which the paradox of the liar is stated, on the one hand, and the technical sophistication required to understand, let alone make progress on, its resolution, on the other. It is not incumbent on everyone to try to resolve every puzzle raised surrounding concepts that they use. Not everyone wondering when the next bus is coming needs to have a view on how to resolve Zeno's paradox that time and change are impossible.

There are other consequences of not being clear about the basic concepts we are using as biologists: we misuse them. We think we have provided explanations when

---

* For reasons that are explored below, this is particularly true of the definitions used in population genetics.

we have not or, perhaps worse, we think we are in a position to deploy the concepts when we are not. It can also lead to our misunderstanding of the sort of evidence we need to be looking toward to test our theories.

The hallmark of science, something that separates it from other intellectual endeavors, is the vulnerability of our scientific theorizing to the evidence the world provides. One obvious fact about evolutionary theorizing is that it is on the one hand easy, and on the other hand, hard. It is easy to come up with explanations and it is hard to test them. This needs to be recognized and explained. What is it about what we are doing as evolutionary scientists that explains this situation? Is it a feature of the world and our relation to it? Or is it that this is an artifact of our particular assumptions in dealing with our subject matter? Anyone who has worked in evolutionary biology has felt the tension between these two aspects. This naturally leads to the question of the scientific status of evolutionary biology. There is a well-worn path from a perhaps idealized picture of the relationship of scientific theorizing to empirical evidence that leads to a paradoxical rejection of the scientific status of most of what we think of as science. Rutherford is commonly credited with an early statement of this "muscular" conception of science when he said "There is physics and there is stamp-collecting (Crane and Mellor, 1995, p. 68)." There may well be something that marks off physics (or at least some bits of physics) from other disciplines in terms of the security of its empirical basis.

As we have seen, the consequences of the circularity argument are more serious than some have given it credit. We cannot acquiesce while it stands because it threatens to make a sham out of our using the theory of natural selection to explain evolutionary change. Natural selection would no longer be a mechanism of evolutionary change. Evolution by natural selection would simply be evolution under a different guise, and natural selection would no longer be an explanation of evolution. The only way to see whether we can escape this conclusion is to see whether the premises are forced on us. In fact, I want to show that both premises of the circularity argument are unacceptable. When we give more appropriate definitions of fitness and of natural selection, we see that the circularity argument does not go through. The lessons are that we gain our explanatory theory and we get much greater clarity about the key notions of fitness and natural selection.

REFERENCES

Burchfield, R. W. 2000. *Fowler's Modern English Usage*, Oxford: Oxford University Press.

Crane, T. and D. H. Mellor. 1995. There is no question of physicalism. In *Contemporary Materialism*, edited by P. Moser and J. D. Trout. New York: Routledge, 1995, pp. 68–94.

Darwin, C. and A. R. Wallace. 1858. On the tendency of species to form varieties; and on the perpetuation of varieties and species by natural means of selection. *Journal of the Proceedings of the Linnean Society, Zoology* 3: 45–62.

Darwin, C. 1859. *On the Origin of Species*. 1st ed. London: John Murray; 2nd ed., 1860; 3rd ed., 1861; 4th ed., 1866; 5th ed., 1869; 6th ed., 1872; 6th ed., with additions and corrections, 1876.

Peirce, C. S. 1877. "The fixation of belief." In J. Buchler, *Philosophical Writings of Peirce*, Dover Publishing; New York, 1955.

Peters, R. H. 1976. Tautology in Evolution and Ecology. *The American Naturalist*, 110: 1–12.

Sober, E. 1993. *Philosophy of Biology*. New York: Oxford University Press. pp. 69–73.

Waddington, C. 1960. "Evolution After Darwin" in S. Tax ed., *The Evolution of Life*, University of Chicago Press, Chicago, 1: 381–402.

3 Resolving the Problem of Circularity

The circularity argument threatens to strip the theory of natural selection of its explanatory power. The theory would be useless in explaining the dynamics of the world because literally nothing is conveyed by adding "by natural selection" to the claim that has evolved. The various versions of the circularity argument are certainly valid; if the premises are true, then the conclusion must also be true. But are the premises true? Were the premises of the circularity argument true and unavoidable, it would mean that Darwin's theory would not be able to play the role he sought for it in explaining both the variety of organic life and its manifest adaptedness to its environments. This is a very serious threat to the theory, which plays such a central role in our understanding of the world. This is the worry to which we turn our attention. In this chapter, we find ways to define the key concepts of fitness and natural selection, which respect the aims Darwin had for his theory. Moreover, there is some reason to think that the way we define the concepts is very much in keeping with Darwin's own understanding and intentions. If this is right, it would go to show that Darwin was very sensitive to the way the theory of natural selection figured in explanations of the biological world. His presentation of the theory, his arguments and explanations, all point toward a solution to the objection presented by the circularity argument.

One natural response is to think that the first premise defines fitness in the wrong way. Rather than defining fitness in terms of survival, what if we defined it in terms of the degree of adaptation? That sounds much closer to Darwin's way of thinking.

Then the argument would be as follows:

1. The fittest are, by definition, those that are best adapted.
2. Natural selection is, by definition, the survival of the fittest.
   Therefore, by substitution of definitions, the following is derived:
3. Natural selection is the survival of those that are best adapted.

This sounds much more tolerable and does not sound at all tautological. In this conclusion, there is a link made between fitness and adaptation, and the second premise links fitness with natural selection. Moreover, the argument does not seem to be near the absurdity of the simple circularity argument. So far, so good, but what do we mean by "best adapted"? Measuring the degree of overall adaptation is very difficult. We certainly do not want to leave the definition of "adaptation" hanging: Justice Stewart's "I know it when I see it" remark would be unacceptable a scientific endeavor (*Jacobellis v. Ohio*, 378 US 184). Remember, adaptation was one of the phenomena Darwin wanted to explain. We could make the assumption that some particular characteristic is the key criterion for adaptation, for example, the number of offspring or the degree of camouflage in some particular environment. That assumption is often expressed using the phrase *ceteris paribus* (all other things being equal). That is, other things being equal, the degree of camouflage determines the degree of adaptation. However, we know well enough that all other things are not equal. We might be able to count offspring, but there will be many characteristics on which the types of organism vary. For example, the rate at which offspring survive to reproduce may not simply be determined by the number of offspring produced. Assuming that none of the other characteristics are important does not show that they are not important. This route to defining the degree of adaptation looks very difficult to negotiate. One thing we had better not mean is that by definition, the varieties that survive are the best adapted. If we were to make such a definition of the best adapted, then we can extend the circularity argument and find ourselves in conceptual trouble again, as follows:

1.1 The fittest are, by definition, those that are best adapted.
1.2 The best adapted are, by definition, those that survive.
    Therefore, from premises 1.1 and 1.2, by substitution of definitions, the following are derived:
1.3 The fittest are those that survive.
2. Natural selection is, by definition, the survival of the fittest.
    Therefore, from premises 1.3 and 2, the following is derived:
3. Natural selection is the survival of those that survive.

We are back in the trouble the original pair of definitions landed us in. Natural selection does not seem to be the sort of concept that can play any sort of explanatory role. The lesson seems to be that the detour through adaptation did not give us the solution to the circularity argument when the degree of adaptation was operationalized using the very thing we were looking to explain: actual survival rates. Again, the problem is not simply that we can derive this conclusion that looks trite and uninformative but rather that, by definition, nonselective evolution is impossible.

We should go back to the simple circularity argument and wonder about the premises directly. Can we find a way of defining fitness not based on actual survival rates? If we can do that, we can reject the premise that defines the fittest as the survivors. Or indeed, are there solid grounds to say that natural selection is not to be defined as the survival of the fittest? If so, then we can reject premise 2?

WHAT IS FITNESS?

Oddly enough, there is an observation in the King James version of the Bible that speaks directly to this concern:

> I returned, and saw under the sun, that the race is not to the swift, nor the battle to the strong, neither yet bread to the wise, nor yet riches to men of understanding, nor yet favour to men of skill; but time and chance happeneth to them all. (Ecclesiastes 9:11)

This observation contains lessons for us. It is certainly true that what the ancient writer was listing does indeed occur. It would be surprising if the biblical writer's claim that the race is not to the swift was intended to convey that the race was never to the swift. More likely, the intended meaning is that the race is not always to the swift, the battle is not always to the strong, and so on. This contention seems right and salutary. It serves as a reminder, as if we needed it, of the vagaries and injustices of life. Our understanding of swiftness is not operationally defined by the actual winning of races. Our understanding of strength is not defined by winning battles. Moreover, our understanding of wisdom and understanding itself is not certainly defined in terms of riches and wealth. These notions—wisdom, understanding, strength, and swiftness—might allow us to explain why the strong win battles, the swift often win races, and those with understanding win riches, but it is also consistent with these notions not eventuating. The sneaking suggestion might be that we could lengthen the biblical list with the observation that survival is not always to the fittest.

The simple circularity argument has just two premises, and if the premises were true, the conclusion would have to be too. If we want to avoid the conclusion, we need to be able to reject at least one of the premises. The first premise says that the fittest are defined as those that survive. In a more contemporary way of speaking, the fittest are not defined by mere survival but by their having a higher rate of survival, that is, they differentially survive. Evolution itself is the change in the composition of the biota, and when there is change in the biota, there must be differential survival; that is the only way to get a change in the composition of the biota. However, if there is evolution, then there is differential survival. In that case, the first premise itself entails that any evolution involves the survival of the fittest. However, most people who endorse the theory of evolution by natural selection do not want to count all possible examples of evolution as selective evolution. Freak events can certainly have consequences for the makeup of the biota. Such events bring about evolutionary change and do not all seem to be examples of natural selection. Darwin's work is riddled with passages denying that natural selection is the only mechanism of evolution. For example, the last sentence of the introduction to every edition of *On the Origin of Species* reads, "Furthermore, I am convinced that Natural Selection has been the most important, but not the exclusive, means of modification."

If natural selection is, in Darwin's words, a "means of modification," then the characterizing natural selection as differential survival has mistakenly focused on the modification and not its means. This confuses the process, natural selection, with the product, evolutionary change. Natural selection is a way of changing the biological world; the change itself is evolution. According to Darwin, there are other actual and possible ways that evolution can come about. Confusing the process, natural selection, with its product, evolutionary change, has the unattractive consequence of making natural selection necessarily coextensive with evolution; there could not be evolution that was not an instance of natural selection.

Darwin and evolutionary theorists in general have not thought of natural selection as the sole mode of evolution. The quoted passage shows that this is right as far as Darwin is concerned, but some evolutionists starting with Wallace himself and persisting with those who wear the mantle "adaptationists" did and do think that natural selection is the sole mode of evolution. However, even those who do think that natural selection is the sole mode of evolution need not think this is a conceptual truth (Wallace, 1914). For example, someone might be an adaptationist, thinking that the biological world is optimally adapted, because they think (as Wallace did think) that natural selection explains all actual organic evolution. Someone who thought along those lines could have also consistently accepted that natural selection might not have explained all organic evolution, that the biological world need not have been the way it actually is. Even that mere possibility is ruled out by the premises of the simple circularity argument. It is a consequence of the premises of the simple circularity argument that it is a nonempirical matter that natural selection is the exclusive mode of evolution. If a trait comes to be characteristic of a population, then its bearers have survived and so by definition must be fitter, and because they survived, they were selected, so their coming to predominate is an instance of natural selection.

The fact that the belief in a plurality of modes of evolution is commonly held among evolutionists merely shows that the premises of the simple circularity argument and this commonly held belief are inconsistent. Leaving the diagnosis of the situation in that state is unsatisfactory because it may be that the premises of the simple circularity argument really are unavoidable for anyone who thinks Darwin's theory of natural selection is true. To show that Darwin's theory can be put into a form that avoids the circularity argument, we have to do more than reject the conclusion. The point is to show where, in the premises, the fault lies and how a properly articulated version of Darwin's theory manages to explain what Darwin sought to explain.

What we have to do now is closely examine the premises of the circularity argument to see whether either is false. Perhaps surprisingly, I will argue that both premises should be regarded as false in the reconstruction of the theory of natural selection.

The first premise says that the fittest are by definition those that survive or, in the locution used by population geneticists, differentially survive. In what follows, some of what I have to say is somewhat critical of the way the discipline of population genetics defines key notions in evolution. Some of the problems could be avoided by presenting the concepts of evolutionary theory in an alternative manner. For example, it is common to see natural selection defined as a directional change of allele frequencies. This defines natural selection to be the changes themselves not the process that brings that change about. The key thing to note is that changes in gene frequencies are the result of evolution by natural selection, so we need to have an account of what processes that lead to changes in gene frequencies are the processes properly identified as "natural selection." The changes in the gene frequencies are just shadows; they are records of the changes in the biota. The shadow is not driving the picture rather than the causal processes that drive the changes. This is why applying the degree of fitness to a genotype is, although as a general rule mistaken, admissible as a heuristic.* However, the first thing to note is that this approach seems to run counter to the intuitive conception we have of the fittest and ways in which we use the notion to explain the world using Darwin's theory.

Imagine an island with two varieties of rabbit. One variety is short haired. They are robust and hard to catch for predators, they breed like the proverb suggests and they get up early in the morning bright eyed and raring to go. The other variety is long haired.

---

* See also Lewontin (1974), although his recognition of this distinction is not wholehearted.

These tend to be frail and easy to catch for predators. They breed fitfully and barely enough to keep their numbers stable. They tend to sleep much of the time and eat poorly. It seems plausible that the short-haired rabbits are the fittest variety. Whether they are the fittest or not, it seems plausible that they could well be. Now suppose that there is a meteor shower that "by chance" sends down shards of rock, which, again purely "by chance," misses all the long-haired rabbits and kills all the short-haired rabbits. As unlikely as this scenario is, it is surely not impossible. Hence, our concepts have to allow for it. In that case, the intuitive thing to say is that the fittest variety was, by chance, the variety that did not survive; this is a case of the least fit surviving.

If we were to use the definitions in the simple circularity argument, however, this description cannot be possible. The long-haired rabbits survived, so by definition of fitness, they are the fittest. Because they did survive, the fittest survived. Then by definition of natural selection, the evolutionary event caused by the meteor shower was an example of natural selection. Something has gone seriously wrong. What looks like a possible scenario, in which the fittest do not survive and the least fit do, turns out to be logically impossible.

If the scenario described where the fittest are in fact, by mere chance, the unlucky ones that fail to survive is to be possible, we must reject the definition of fitness used in the circularity argument. The definition in premise 1 fails to explicate the intuitive concept of the fittest because it rules out the intuitive possibility of the fittest not being the ones that survive.

At this point, the definition of the fittest becomes the issue of some importance. If the fittest are not by definition the individuals that in fact survive, which individuals are they? A small step forward can be taken when we note that the fittest individuals are the best adapted individuals, but the same question arises with respect to the notion of the degree of adaptation. If the notion of fitness is conceptually tied to that of adaptation, we need to shed whatever light we can on both these notions.

FITNESS AND ADAPTATION: THE KEY CONCEPTS

Fitness and degree of adaptation are relativized to environment. An ocean going pelagic fish such as a bluefin tuna is not well adapted to life in shallow, murky reed beds, let alone life on land. Albatrosses are well adapted to a life soaring on the oceans and not well adapted to tropical forests. This is agreed on all hands, even if it is far from clear what is going to count as the environment. At least a part of the problem is the extent of the environment. For example, the environment contains spasmodic random and catastrophic events such as earthquakes, meteor showers, severe fires, and the like. These events are part of the environment when they occur, but should they be regarded as exercising selective pressure on organisms or as mere stochastic effects? On the other hand, change inevitably occurs. Is change a part of the environment, or is it the supplanting of one environment with another? The answer to these questions is far from clear, but what is clear is that the answers will not be general. In some circumstances, a very infrequent catastrophic event such as severe wildfire can lead to changes in the fauna and, in particular, the flora in a region. This is certainly the case with many *Acacia* species in Eastern Australia, the seeds of many of which require exposure to fire before they germinate.

For this reason, there is often a strong sense of the ad hoc in the characterization of the environment relative to which fitness is defined. This has led some to doubt whether comparisons across species or varieties with respect to fitness or adaptation even makes sense (Lewontin, 1978, 1985). I disagree. Leaving aside empirical questions of measurement and so on, one of the two key phenomena the theory of evolution by natural selection is required to explain is the manifest adaptation of organisms to their environment. It is just because it is so obvious that organisms are so overwhelmingly well adapted that an explanation is required, whatever form such an explanation is to take. To be clear, adaptation is something that can be observed. Organisms and their environments are not randomly sorted. Organisms tend to occur in environments for which their features provide help. Darwin's theory was, as we have already seen, aimed at explaining that manifest fact. It is important to realize that it is conceivable that the biological world was not organized like that. On the one hand, we can imagine that the very same organisms occur across a wide range of environments. On the other hand, we can conceive of situations in which organisms occur in environments to which they are not suited. This is conceivable but not what we observe.

When we try to become more precise about what we mean by the degree of adaptation, we come across very thorny and seemingly intractable problems. How, for example, should we begin to define what we mean by the environment to which the organism is adapted? There is a risk that in letting the organism define its environmental niche, we will be forced into conceding the absurd consequence that all organisms are equally well adapted to their own environment because, by definition, they fit it perfectly (Peters, 1976). The theory of natural selection requires that there be variation in fitness among varieties. If this is a conceptual impossibility, then the theory will founder.

Darwin himself did not actually argue that he had unequivocal evidence that some varieties are better adapted than others. What he says is that if some variety has a beneficial modification, then it will come to dominate. Moreover, Darwin (1859, especially in chapter 1) argues that because variation seems undirected and that there is so much variation of so many heritable characteristics, it is plausible that some of these variations are beneficial to the organisms.

It is because of their adaptations that organisms are adapted to their environment. Because natural selection has been said to explain adaptations, some theorists have sought to define the notion of an adaptation in terms of natural selection. This approach has been called the selective account of adaptations and has been particularly applied to the related notion of biological function. The function of a trait on this understanding is the reason for which it was selected. This account was articulated by Larry Wright (1973), and this sort of account, as Karen Neander (1991a, 168) reported, is the consensus view. This selectionist approach also gets called "the etiological concept of function." However, because the name "etiological" comes from "etiology" from study of causes and causes are both up- and downstream of the trait's presence, the term "etiological" is misleading. I will not be using that term. This view may well be a consensus, but it is important to see that this selectionist account is mistaken.

Eliot Sober is prominent among those who define an adaptation as a feature that has its origin in natural selection. Karen Neander (1991b), defining the related notion of a biological function, characterizes the function of a feature as the causal consequence of the feature that led to its selection. In both characterizations, the defining criterion for a feature to be an adaptation is that it has its origin in natural selection. Each of these seemingly innocuous characterizations has unfortunate consequences for the theory of evolution by natural selection. This means that we cannot discover whether a feature is an adaptation without knowing its history. However, it is important to remember that the theory was supposed to explain something we can understand without having a view on the history of the origin of features; their adaptedness to their environment cannot even be recognized except in the light of the theory of natural selection. That is, that a feature is an adaptation is not a relationship between the organism the feature is a feature of and its environment but rather is a feature of the history of that feature. However, this attempt at a definition of the notion of adaptation is not compulsory and makes little sense of the following sort of scenario.

Suppose, what seems plausible, that some cataclysmic event such as a massive asteroid or comet strike occurred at the end of the Cretaceous and led to a massive change in the biota including the extinction of the dinosaurs. The fact that some Cretaceous mammals such as *Alphadon*, a metatherian, survived may well have been because of their homeothermy, their ability to regulate their body temperature in a physiologically active manner.* This may have made them less reliant on ambient temperatures and so better able to survive.† However, this would not mean that they were homeotherms because of the selective pressure raised by the rapid cooling of the Earth. The homeothermy could be adaptive in that environment without its origin being due to that selective pressure. For that reason, it would be mistaken to simply equate the reason a trait is an adaptation with the trait's origin. There are many philosophers of biology who have argued for this position.‡ Apart from the fact that it fits ill with the intuitive picture we have of the notion of adaptation as shown by the examples, there is a more important reason for refusing to define adaptation in terms of natural selection. This more important reason is that adaptation is one of the two phenomena Darwin set his theory to explain along with the other phenomenon, biological diversity. We can understand both of these phenomena without a prior commitment to Darwin's theory as an account of the origin of the phenomena. We understand, just as Paley (1836) and other pre-Darwinians understood, that there is a diversity of organisms. If Darwin was right in his theory of natural selection,

---

* This is an importantly different notion from the notion of warm-bloodedness. Not all warm-blooded organisms are homeotherms.

† In fact, for all its plausibility and frequent repetition, we ought to be skeptical about the initial plausibility of this sort of story, which I am using for illustrative purposes. The work of Tom Rich of the Museum of Victoria and Pat Vickers-Rich of Monash University and their teams has thrown a fly into this particular ointment (Benson et al., 2012; Woodward et al., 2011). They have discovered polar dinosaurs, which would have endured significant periods of total darkness and cold even before the cataclysmic closing of the Mesozoic. I should also note that crocodilians, which are reptilian and not homeotherms, do maintain a fairly constant body temperature although they spend a considerable amount of their time in water, which absorbs heat at a much faster rate than air of the same temperature would. Their maintaining a constant temperature may have a lot to do with their restricted range, mostly within the tropics.

‡ Among them are Karen Neander (1991a,b), Millikan (1984, 1989a,b), Bigelow and Pargetter (1987), and Sober (1984).

he would have supplied the explanation of the origin of that diversity, but he would not have supplied, with his theory, the definition of that diversity. Conceiving of the diversity simply does not depend on conceiving of natural selection. Natural selection is the cause of the diversity, not the definition of the diversity. The point is being belabored because exactly the same situation arises with being adapted, and as we have seen, many people have failed to see how Darwin's theory relates to the phenomenon.

Darwin wanted his theory to explain how it was that organisms are adapted to their environments, and he thought that a big part of that is explaining the origin of adaptations that adapt organisms to their environments. Stand back from the modern view for a moment. Let yourself adopt the view of a religious person, perhaps one who has heard the lectures on natural theology given by Reverend Thomas Paley, whose work Darwin thought so highly of as a Cambridge student. Paley argued for the existence of God from the adaptations that suit organisms to their ways of life. Paley then sought to explain the very thing that Darwin sought to explain. Darwin gave one sort of explanation; Paley gave his theistic explanation. However, they both agreed about what they sought to explain.

Organisms are adapted to their environment to the same degree that they are, regardless of whether they evolved by natural selection alone; evolved by natural selection together with other mechanisms of evolution; were created by an omnipotent, omniscient God; were created by a less than omniscient and omnipotent God; or indeed popped into existence ten minutes ago as a bizarre but not impossible quantum singularity. Their degree of adaptation is a matter of their relationship to their environment, not a matter of how they got to be in that relationship to their environment. In that case, any definition of the notion of an adaptation as a feature that is in any way conceptually tied to natural selection must be mistaken. We can still identify adaptations by relating them to their impact on the degree of adaptation of the organism. However, the challenge is to characterize the degree of adaptation without presupposing the theory of natural selection. To complete this reformulation of Darwin's theory is to spell out adaptation in terms that do not require an antecedent understanding or commitment to natural selection.

Sober, who does think that to describe something as an adaptation is to refer to its origin through natural selection, provides an argument for distinguishing between a trait being an adaptation and a trait being adaptive. He asks us to consider granting that wings in a species of bird are adaptive and adaptations for flying. Now he asks us to consider what happens when flying becomes positively dangerous because of a new flying superpredator, to the point where flying is not adaptive. Then he suggests wings are still adaptations but are not adaptive. His description of that case should be resisted because it appears to be quite misleading.

Several points need to be made: whether some traits are adaptive, and therefore whether an adaptation is relativized to environment. So what is true of the wings is that they were adaptive in one environment and may not be in the new environment containing the flying superpredator. What we would say if we felt that the wings were no longer adaptive is that they were adaptations but are no longer. Or we would say that in such and such an environment, they are adaptations, but in so and so an environment, they are not.

Thus, the consequences of the costs and benefits are clear; we need to relativize what is an adaptation to an environment. However, let us be clear that even so, the second point to notice is that it is far from obvious that in the new environment, the wings are not adaptations and hence adaptive for the birds. Suppose that flying is positively dangerous in the new environment with a flying superpredator. That is, suppose that flying is much more risky than staying on the ground. Still one needs to know what is adaptive for the individuals of this bird species. It may be that if you had a choice between being a sparrow and a snail in this environment, you should choose to be a snail, but that is not a choice that confronts any individual of the sparrow species. It is stuck with wings, a beak that may be adapted for feeding in particular ways, and so on. The choice presented may be between running a big risk of predation by flying and running an even a bigger risk of starving to death by staying on the ground. In that case, although flying is much riskier than it was, it is for that species still adaptive, and the particular wings it has are adaptations for flying.

Biologists can represent a population in a model as subspace of a multidimensional hyperspace, with one of the axes measuring fitness and other axes indicating significant environmental, behavioral, phenotypic, or genotypic parameters. This technique was used originally by Sewall Wright in evolutionary theory and by G.E. Hutchinson in ecology. The hyperspace is often called a "fitness landscape" where height on the fitness axis is taken as primary for the landscape. A point in the landscape where any small change in the parameters measured would lead to a drop in fitness is a local maximum and is represented on the fitness landscape as a mountain top. In the case considered by Sober, it may be that the flying ecological niche is lower than many nonflying ecological niches on the landscape. However, given the shape of the fitness landscape, it may be that immediately surrounding the local maximum where the flying species is located on the landscape are areas much lower in fitness than the local maximum it inhabits. In that case, for that species in that new environment, any move to a nonflying ecological niche involves a step down in fitness. The fact that there are nonflying niches in principle available, which have a much higher fitness associated with them, is utterly irrelevant from a selective point of view. The move to these niches may involve a big evolutionary change and for that reason may not be feasible. Natural selection cannot drive changes that take a population through a drop in fitness.

There is a further important lesson in this apart from showing us that we should not define adaptation in terms of selection. The notion of an adaptation is not an absolute one even relative to the environment. The organism is sometimes suggested to be the best overall solution to the prevailing selective pressures. That is, it will be at the global maximum on the fitness landscape. It is much more likely to be not the best overall solution but the best of the alternatives available. That is, the landscape we consider must be partial and does not cover all possible combinations. This qualification can hardly be overemphasized, but it is all too often disregarded. Because all we really know, even when we know that a trait was directly selected, is that it made the organisms with the trait better adapted than the other varieties then in existence; we cannot jump to the conclusion that the organism thus selected is the best overall compromise because not all alternatives have competed. Moreover, such a supposition would abstract away from frequency-dependent features of characteristics. It is a well-known phenomenon that the fitness of a characteristic may depend on the relative frequency of it and other characteristics in the environment. At this point, it is pertinent to note that natural selection is not the only mode of evolution, so it may also be that a fitter variety did exist but fell by the wayside for nonselective reasons. Then it is true to say that natural selection, acting on the remaining varieties, led to what we see today. The outcome has been selected but not simply selected: Other modes of evolution have played a role in bringing it about.

A biological example, from the application of game theory to evolutionary modeling, indicates just how readily the scope of selection is disregarded. The notion of an evolutionarily stable strategy (ESS), developed by John Maynard Smith (1976, 1979, 1983a,b), is used in theoretical population genetics. Maynard Smith (1976) introduces the notion as follows:

> If a particular strategy, say I, is to be an ESS, it must have the following property. A population of individuals playing I must be "protected" against invasion by any mutant strategy, say J. That is, when I is common, it must be fitter than any mutant.

As the notion is defined, an ESS must be such that no mutant version could take over. This definition requires that strategy is a global maximum. After all, were the strategy in question just a local maximum or even just the best of the actual competing variants, it would be selected but would not be "protected against invasion by any mutant strategy." If a fitter variant enters the competitive space, that variant will be selected. For that very reason, there seems little reason to believe that we have ever seen an ESS, or that we could recognize one if we did. To recognize an ESS, we need to know the character of the complete fitness landscape and not simply know the fitness of the existing and competing varieties. In any case, it is safe to assert that not one adaptation has ever evolved through natural selection weeding out all possible alternatives because not all possible alternatives have competed. At any point in the world's evolutionary history only finitely many varieties have constituted the selective space, a very small subset of the total varieties possible.

A feature is adaptive if it is a "good enough" or satisfying solution to a problem faced by the organism, which is part of a "good enough" package of the alternatives available. The fact that there is a better overall solution on the fitness landscape, but one that is not available, should raise no tendency for us to regard what we have before us as less than adaptive. Adaptive does not mean maximally adaptive, nor does it mean better than all the other varieties around. It is important to note that there are numerous senses in which what we see in the world may be regarded as less than optimal. First, to use the metaphor of the fitness landscape, the peak on which the population is to be found may not be the highest one. Second, a population is not a point in the landscape but rather a region. That region may not include the local peak, or the distribution of the population may include the local peak but some less optimal regions as well. Organisms that are short of being even on a local peak in the fitness landscape are never the less more adapted than could have been.

These points allow us to see some of the problems with the sort of adaptationist strategy endorsed by Daniel Dennett (1995) in his book *Darwin's Dangerous Idea*. Dennett gives a typically spirited defense of the idea that what we have before us in the biological world should be regarded as the overall best compromise given structural constraints and so on. The first thing to point out about this claim is that it has the implausible consequence that we happen to live at a time at which there is no directional selective evolution.

Because what we have before us is the best overall solution to the environment's demands, there is no scope for change through selective evolution. This makes no biological sense: why should we be at this unique point in biological history when populations are no longer changing through directional selective evolution? There is an even more important reason, however, to reject the view that the biological world is optimally adapted. Such a view is inconsistent with one of the features of the biological world the significance of which Darwin was the first to see: variation. Populations vary in many characteristics. Playing down the significance of this fact inevitably leads to a misconception of the theory of natural selection. That variation that Darwin drew attention to had been noticed by others, as Huxley remarked. What then is the claim that organisms are optimal supposed to say about this variation? Are all varieties equally optimal? Is there supposed to be a selective explanation for why this variation exists? Notice that explaining why variation exists in a population in terms of selective advantage is not the same as explaining why the particular variation we see exists.

Having seen the problems we face in giving an account of the theory of natural selection, we are left with the following criteria that the concepts must satisfy:

1. The fittest (best adapted) are not by definition those that survive.
2. There is the possibility of variation in the degree of fitness (degree of adaptation).
3. The theory of natural selection can play a role in explaining the extent of adaptedness we observe in the world.
4. The theory of natural selection can explain the origin of biological diversity.

We have seen that it is a mistake to define something as an adaptation if it has its origin in natural selection. There are all sorts of features that may be nonadaptive or even positively deleterious to the organism, but which have their origin in natural selection. We can see this from an example noted by Darwin.

> From facts collected by Heusinger, it appears that white sheep and pigs are injured by certain plants, whilst dark-coloured individuals escape: Professor Wyman has recently communicated to me a good illustration of this fact; on asking some farmers in Virginia how it was that all their pigs were black, they informed him that the pigs ate the paint-root (Lachnanthes), which coloured their bones pink, and which caused the hoofs of all but the black varieties to drop off; and one of the "crackers" (i.e. Virginia squatters) added, "we select the black members of a litter for raising, as they alone have a good chance of living."

This is a fascinating passage and was inserted by Darwin for the third edition, expanding a much shorter and cursory comment in the first two editions. The passage thereafter remains unchanged despite other revisions in the many editions leading up to the final edition of *On the Origin of Species* in 1876. Imagine now that, instead of the artificial selection by the farmers, a population of pigs underwent natural selection in a context where the paint-root was a major food source. Clearly, the ability to detoxify the paint-root would be selected, and more of the variety that had that ability would survive to reproduce. In addition to that characteristic coming to predominate in the population, the color of the pig population would also change from being various colors to uniformly black. The black coloration is not itself advantageous; in fact, it might make the pigs easier for their predators to see, but the change in the population of pigs from a multicolored population to one that is characteristically black would have had its origin in natural selection, but not because being black is adaptive. Therefore, not every feature that has its origin in natural selection is adaptive, and it is also true that not everything that is adaptive has its origin in natural selection. To be clear, the pigs that are black are better adapted to that particular environment but not because they are black. Rather, they are better adapted because they can break down the toxins in the paint-root. Thus, although the population of pigs evolves and comes to be black through natural selection, we need to be able to identify one aspect of the pigs that comes to predominate but is not being selected for (the color) and another aspect that comes to predominate but is being selected for (the ability to detoxify the roots).

So far, we have seen that it is a mistake to define the fittest as those that survive. It is also a mistake to define the fittest in terms of actual survival numbers; this includes the common accounts of fitness in terms of those that differentially survive. Such an account fails to explain the notion of fitness adequately. This much is not controversial. What is controversial is the reason for the inadequacy of such an account of fitness. I have tried to suggest the reason that to define fitness or adaptation in terms of actual survival would strip the theory of natural selection of its explanatory power.

Fitness or degree of adaptation is the key notion in the theory of natural selection. This may seem odd because natural selection is supposed to explain adaptation and yet on my account of the theory presupposes it. This is quite correct. The theory of natural selection does presuppose the notion of fitness, or adaptation, and does explain it as well. However, what gets presupposed and what gets explained are distinct. What gets presupposed by the theory of natural selection is that there is heritable variation in characters, which ground differences in fitness within a population. What gets explained is how the population will come to be represented over time by the more fit individuals. Thus, the presence of adapted individuals in the population is explained in terms of the way that population varies and the consequences of that variation in a competitive environment. Thus, fitness is both presupposed and explained by the theory of evolution by natural selection.

FITNESS AND PROBABILITY

In the simple circularity argument, fitness was defined as follows:

1. The fittest are, by definition, those that survive.

The problem with this definition is that it does not leave room for the situations in which the fittest, for one reason or another, are not the ones that survive. What we need is a way of defining fitness that does not have the consequence that the fittest are always by definition the survivors.

We might think that although fitness should not be defined in terms of actual survival numbers, survival numbers are relevant: survival numbers are evidence of fitness. However, even if survival is recognized as evidence for fitness, it is not conclusive evidence. Survival can be impinged upon by many factors of which fitness is just one. One way of modifying the premise of the circularity argument to take account of the fact that the fittest may not survive is to define the fittest as those with the greatest probability of surviving. The thought is we replace premise 1 with an alternative, as follows:

1.3 The fittest = (by definition) those with the greatest probability of survival.

The thought is that although the fittest may not be the ones that survive in particular circumstances, the fittest do have the highest probability of survival. By analogy, suppose we had a die which had five black faces and one white. That the die would come up with a black face has the greatest probability (assuming the die was fair). It is not impossible that on rolling the die just once, we find it is the white face that comes up. Similarly, there is no contradiction in supposing that the individuals with the greatest probability of surviving in fact do not survive. Moreover, we can see why actual survival numbers count as defeasible evidence, but evidence no less, of the degree of fitness. We quite rationally tend to think of a coin that turns up heads every time as biased. Thus, if one variety survives differentially, then we tend to think of it as the fittest. In the opening scene in Tom Stoppard's (1967) play *Rosencrantz and Guildenstern Are Dead*, we see Rosencrantz and Guildenstern engaged in a gambling game involving a coin. Rosencrantz is betting heads on each coin toss and wins ninety-two bets in a row. Guildenstern does not doubt the fairness of the coin but reaches for other explanations. This scene makes play with the absurdity of treating every sequence of heads and tails as equally evidential of fairness in a coin. It remains true that every sequence of heads and tails is equally logically consistent with every nondogmatic assignment—that is, assignments that do not assign probability 0 or 1—of a probability of a heads coming up.

Although the introduction of probability into the definition of fitness seems very promising, the nature of this probability, or in Darwin's words "chance," is a pressing matter. I want to show that none of the standard interpretations of the notion really fit what we need for this theory and that the only way to really understand the use of the notion of fitness in evolutionary theory is to replace the probabilistic suggestion with another sort of account.

Intuitively, the idea that the fittest are the most likely to survive makes sense and seems attractive. The thought is that if we had a die with five black sides and one white side, then it is most likely that a black side would come up, but it is possible if we roll that die just once, wherein on that one roll, a white turns up. That seems to fit the bill. Black is most likely to come up, but it does not. Similarly, we want to be able to say that the fittest individuals are the most likely to survive, but if things go ways that they could go, all the fittest individuals do not survive. So far, so good. However, the question is, What do we mean when we say that the most likely outcome does not occur? That forces us to think about the nature of probability, something we can often pass over in silence. Focusing on that question here is forced on us.

There are several interpretations available of the notion of probability (Hájek, 2012):

1. Long-term average frequency
2. Subjective betting rate or partial belief
3. Logical probability
4. Objective chance or propensity

What we will do now is see whether there is a way of understanding the idea that the fittest can be defined as those that have the greatest probability of surviving. What we need is to see whether one or more than one of these accounts of probability will allow us to make sense of the way that the fittest need not be the ones that actually do survive. After all, I could have a fair die with five red sides and one black, and although it seems that red has the highest probability of coming up, when I do come to throw it for the one and only time it will be thrown, it is black that comes up. In that case, the color that was less likely to come up nevertheless comes up.

What do we mean by "probably" in 1*? This is just not clear, and this is because there are many notions of probability. That would not be a problem if there were any notion of probability that allows for the notion of fitness to satisfy the conditions we need for the theory of natural selection.

LONG-RUN FREQUENCY ACCOUNT

One way of understanding probability is as a ratio of cases, the long-run frequency interpretation. When we have a type of event, say that tossing of a particular coin, we can compare the relative frequency of the possible outcomes. We start off with very few coin tosses. The number of heads divided by the number of tosses gives us a ratio. As the number of tosses gets bigger and bigger, each toss has less and less influence on the relative frequency of heads. Thus, it seems as the number of tosses approaches infinity, we should find a limit, a long-run frequency to which the particular system we are interested in tends.

According to the long-run frequency account of probability, strictly speaking, a single event can have no probability assigned to it as a single event. We can assign an event a probability when it is considered as an exemplar of some general property possessed by many events. Any one event will exemplify many general properties, and so the same event will be an instance of many types. Each of these types will have a long-run frequency associated with it. Therefore, the same event in this way gets assigned many different probabilities as an example of many different types of event. Thus, in the hackneyed example, a toss of a coin, the single event of my tossing a coin has no probability of turning out heads as such but only under descriptions, such as the tossing of a coin with such and such physical characters, the tossing of a coin by a man, the tossing of a coin on a Tuesday in a strong Easterly wind, and so on. Not all of these will yield the same probability that the toss will yield a head. The frequency of heads turning up when the coin is in fact tossed on a Tuesday in a strong Easterly wind need not be the same as the frequency of heads tossed when a coin is tossed by a man.

The idea of defining fitness in terms or long-run frequencies of survival numbers might seem initially very attractive. After all, we saw that survival is defensibly but still actually evidence of fitness. Hence, in the long run, one might imagine that the fittest will survive. Moreover, because we are in fact dealing with significant numbers and with long runs in evolution, perhaps the long-run frequency account will capture the sense in which the fittest are those with the greatest probability of survival. Unfortunately, this is unlikely to work out so smoothly. Although the numbers of individuals may be large, it is unlikely that the long-run average frequency notion will answer to the demands of biologists, and this is for the straightforward reason that long-run average frequency is determined by the actual frequencies. In this case, it means that the probability of survival will be totally determined by actual frequencies of survival. The point of the appeal to probability in the theory of natural selection was to allow the talk of cases in which the fittest individuals do not in fact have the highest rate of survival. We wanted to define the fittest as those with the greatest probability of survival to allow for the case where the fittest do not survive. Those were the cases in which we wanted to be able to say that the individuals with the greatest probability of survival did not in fact survive. However, in just that sort of case, the fittest will not have the highest long-run average frequency of survival. Think of the die with the five black faces and the one white face. We wanted to say that the probability of a toss yielding a black face was greater than the probability of a toss yielding a white face. Suppose that die is tossed just once and in that one toss the white face comes up. By definition, according to the long-run frequency account of probability, the side with the greatest frequency of coming up is the white face. Thus, that die has the highest probability of coming up white (not black) according to this account of probability in that situation. The notion of probability as long-run average frequency is not adequate to the purpose of explicating the notion of fitness.

Once again, we see the dangers of using the actual course of events in the definition of the key theoretical terms of the theory of natural selection but now entering into the account of probability we might have been tempted to use to make sense of the theoretical terms of the theory.

Notice that the objection to this account is not that the long-run frequency account is an incoherent account of probability. It certainly is, and should be accepted to be, a coherent and often useful notion of probability, but it will not provide a notion of probability that will allow us to define the fittest as the ones with the greatest probability of survival. The issue being that we want to be able to accept as possible that although certain individuals did not survive, they were nevertheless the fittest. History being a single run, we cannot say the fittest had the highest probability of surviving in the long run although they did not in fact survive. In the long run, the ones that did survive survived. Thus, by this account, they had the highest probability of surviving. Thus, this notion of probability would not play the theoretical role we were looking for.

SUBJECTIVE BETTING RATE ACCOUNT

The long-run frequency account treated probability as a feature that emerges from a ratio of instances of a given type of event. It depends on there being actual rates of occurrence of some phenomenon in the world. There is another notion of probability that focuses on the notion of degree of belief. This says that to say that an event has the probability of 0.5 according to an agent is to say that the agent will regard a $1 bet that returns $2 as fair. To say that an event has a probability of 0.2 is to say that the agent will regard a $1 bet as fair when it returns $5. This makes an agent's probability judgments coordinate with their judgments of risk and reward. For this reason, this account is known as the subjective account of probability: It focuses on the judgments agents do and do not make. The subjective account of probability regards an assignment of probability to an event as an injunction to believe to a certain degree in the proposition that the event will take place relative to a course of evidence (Ramsey, 1931). A subjective account requires a story of the nature of the relationship between the evidence and the strength of the partial belief. The subjective theory of probability locates probability in the wrong spot for it to play an explanatory role in a theory of the world. There is no doubt it is a well-defined notion, and the subjective theory of probability may well play a role in explaining our behavior. Suppose we believed the world to be deterministic. In that case, we would think that there is a fact of the matter about what will happen with the next toss of a particular coin. If it were a deterministic world, then the toss will be heads or tails. We nevertheless would not have subjective confidence in the coin turning up heads if it was going to be heads. Our confidence in heads turning up will be something significantly below 1.0. Notice that subjectivists do not need to say that your confidence should be 0.5. Our betting rate will march in step with our confidence, and this is exactly what the subjectivist says.

In any case, there is some tendency to balk at the idea that a theory about the world like Darwin's, which apparently trades on the notion of probability of survival as a key notion, is actually talking about how we would form beliefs of various strengths given certain evidence where this has no ground in the way the world is.

The subjectivist account is well placed to explain the rationality of an action by treating that as an internal feature of the agent's cognitive state. Why did the agent accept the bet? Well they were asked $1 for a bet that offered $10 and they thought the probability was 0.2. Such explanations relate the internal cognitive states of the agent.

This very feature of the subjectivist account brings with it a strange consequence for the theory we are exploring. Remember we want a theory that says that fitness is tied to this property of being the most probable survivor. And this theory will be explanatory. We want to be in a position to explain changes in the world by using the theory of natural selection. The subjectivist account puts the explanation in the wrong place. There is nothing in the world correlated with subjective betting rates. Subjective betting rates are aspects of our rationality, and there can be many of them. First, how can there be many fitness? Second, how are these many betting rates going to explain the dynamics in populations Darwin's theory is supposed to explain? It looks like the probability is in the wrong place: in our heads when it should be in the world. The best subjective accounts seem to be able to provide is accounts of our beliefs about the organic world, not an account of something that plays a role in the dynamics of the organic world.

It remains true that the subjective account of probability is a theory of something that can indeed play a role in various theories. It is just that it is not suitable for the dynamics of populations of organisms.

LOGICAL ACCOUNT

Deductive logic has the property that whenever the premises of a valid argument are true, the conclusion must be true. We can see that as the premises giving 100% support to the conclusion. Starting with that idea, many thinkers have thought there must be a generalization of that idea where the premises give less than conclusive support for the conclusion. Perhaps seeing a certain number of black ravens, for example, does not give conclusive support to the theory that all ravens are black, but it must support it to some degree. If we could know what that degree was, then we would be in a position to know how confident we should be about any theory given certain evidence. That was the intuitive idea that motivated this approach. For many of the thinkers who sought such a theory, this would be an objective relationship between the premises and conclusion and would allow us to make sense of the way more information would be evidence for or against the conclusion.

The logical account sought to treat conditional probability, say $P(A/B)$, as the degree to which B supports A. In a logical entailment, should B logically entail A, then $Pr(A/B) = 1$. The aim for the investigators of logical probability is to figure out what happens, in general, when B does not logically entail A, and to do so in a principled *a priori* manner.

When Rudolf Carnap (1950, 1952, 1963) provided his very influential approach to this problem, he began by developing a machinery of "state descriptions." These were ways the world could possibly be. The suggestion was that by listing all the possible ways the world could be, we could determine a ratio that would allow us to determine *a priori* probabilities.

Suppose we consider an example to see how it goes. Suppose we have two properties G and H and two objects b and c. This gives us 16 possible states the world could be in and 5 state descriptions that could correctly describe the world.

States	State Descriptions	m	m*
Gb & Gc & Hb & Hc	all positive	1/16	1/5
~Gb & Gc & Hb & Hc	three positive	1/16	$1/20 = 1/(5 \times 4)$
Gb & ~Gc & Hb & Hc			
Gb & Gc & ~Hb & Hc			
Gb & Gc & Hb & ~Hc			
~Gb & ~Gc & Hb & Hc	two positive	1/16	$1/30 = 1/(5 \times 6)$
~Gb & Gc & ~Hb & Hc			
Gb & ~Gc & ~Hb & Hc			
~Gb & Gc & Hb & ~Hc			
Gb & ~Gc & Hb & ~Hc			
Gb & Gc & ~Hb & ~Hc			
Gb & ~Gc & ~Hb & ~Hc	one positive	1/16	$1/20 = 1/(5 \times 4)$
~Gb & Gc & ~Hb & ~Hc			
~Gb & ~Gc & Hb & ~Hc			
~Gb & ~Gc & ~Hb & Hc			
~Gb & ~Gc & ~Hb & ~Hc	none positive	1/16	1/5

There are 16 states the world could be in, and there are 5 state descriptions that could correctly describe the world. If the states are each equally probable, then each state will have a probability of 1/16. If each of the state descriptions is equally probable, then the probability for each state description that it correctly describes the world is 1/5. For each state description, the probability will be apportioned equally within it to the states meeting that state description.

A: Everything is G. B: Gb.

What is P(A/B)? This is P(A&B)/P(B).

If states are equiprobable, then this will be (1/4)/(1/2) = 1/2.

If the states' descriptions are equiprobable and probability is assigned equally to each state within a description, then this will be (1/3)/(1/2) = 2/3.

We seem to have no reason to think one of these ways deciding the assignments of probability is to be preferred. This suggests that the logical account of probability is very sensitive to ways in which the situation is described, and so it is.

This vulnerability was brought home to me in a discussion many years ago with John Collins of Columbia University who presented a scenario he learned from Bas van Fraassen.* Suppose there was a factory that made cubes. The side length of the cubes varies between 0 and 1 foot. What is the probability that a randomly chosen cube has a side length that is less than or equal to 1/2 foot? If we equally distribute the probabilities over the lengths, the temptation is to say 1/2. However, this very scenario can be described instead in terms of area rather than length. The factory produces cubes with a face area between 0 and 1 square foot. What is the probability that a randomly selected cube has a face area of 1/4? That seems to invite the answer 1/4, and it gets worse. We could describe the same scenario in terms of the volume of the cubes. After all, the cubes vary between 0 and 1 cubic feet. So what is the probability that a randomly selected cube has a volume of 1/8 cubic feet or less? We are naturally led to a new and different answer, yet these are three ways of presenting the very same situation.

The other approaches to probability we have considered are well defined and do genuinely characterize distinct notions of probability that can be useful. The logical account is not in the same category. To say that the notion of logical probability is contested would be an understatement. At this point, we have no reason to think that there is a workable version of the logical approach. In fact, it seems that the problems with logical probability leave us unable to say with any confidence that it is this notion of probability we have in mind when we talk about the fittest having the highest probability of surviving.

PROPENSITY OR OBJECTIVE CHANCE ACCOUNT

The subjective account of probability was unhelpful by putting the probability into the head. It is quite mysterious how an agent having degrees of confidence that something will survive can play an explanatory role. It must be emphasized that it is a central feature of the subjective approach to probability that degrees of confidence

---

* This example is developed from the presentation in van Fraassen (1989), but van Fraassen had been discussing this sort of case long before that book was published. When I got to graduate school and learned directly from Bas van Fraassen, I heard the example from the horse's mouth.

do not pick out or march in step with features of the world. There is no further explanation of the degrees of confidence agents take to events. Moreover, there is no question of an agent being wrong about their degrees of confidence, so long as their assignments meet the formal constraints supplied by the axioms. There are very numerous assignments that will meet those constraints. There are also consistency constraints for updating of changes to an agent's commitments, but we do not need to go into here. Were fitness to be understood in this subjective manner, it would appear difficult to understand how fitness could play a role in explaining how things change in the world. Such an account, focused as it is on agents and their commitments, can explain changes in our beliefs, but not why one sort of rabbit survived and another did not. As my subjective confidence about the coin turning up heads is one thing and the way the world goes is quite another, so it will be with my confidence that a variety survives and whether they actually do survive.

Were there an explanation of the change in rabbit numbers, it would seem it would have to be a feature that both explained why our confidence in their survival was greater and also why it was greater. That could not merely be a feature of our subjective betting rate but rather something that subjective betting rate tracked in the world. In other words, it would be something inconsistent with the subjective account of probability.

We saw that an alternative account of probability, the logical account, promised that there would be a story explaining detailing given some evidence just how much support that evidence gave to a theory. According to the logical account, this is not a matter of how much confidence an agent has in the theory given the evidence but how much confidence they should have in the theory given their evidence. This sort of account, different from the accounts we have explored, is of questionable coherence. The long-run frequency account and the subjectivist account each seem to clearly define a notion of probability that might find theoretical interest for certain purposes. It is just that they will not provide a concept of probability that plays the role we wanted in the definition of fitness. The logical account, however, comes with many promissory notes and none have been delivered.

There is another interpretation of probability that is in fact the one favored by many scientists, which regards probability as a measure of objective chance, that is, as a real feature of the world (Gillies, 2000a,b). This notion was termed the "propensity" account by Karl Popper (1957, 1959). Mills and Beatty (1979) treat fitness as a propensity to survive and therefore as an objective feature of the world. Propensity is supposed to be a real feature of the world. In physics, the idea that there are, at the quantum level, real propensities is now taken very seriously. The idea that two radium atoms may differ in their propensity to split in the next 60 seconds may well be an underlying explanation of the way these objects evolve. Propensities are features of the world, there to be discovered by us or not, like other features of the world. They figure in our explanations of the way things change or not in the world too. We also have an account of what we are doing when we do biology: we are trying to figure out how different characteristics in different environments determine the propensity to survive of their bearers.

This sort of account does indeed have many attractive features; it makes sense of the theory of natural selection as relating to the nature of the world. It apparently salvages, just as the subjectivist account did, the claim that the fittest do not always differentially survive. However, there seems to be a significant problem with the propensity account.

It has the consequence that it makes explanations using the notions of fitness and selection hostage to the final form of physics. If that physics is deterministic, then individuals have no propensities to survive other than 0 and 1. Under deterministic situations the propensity of any particular event to take place is exactly one of 1 and 0. In that situation, when we consider the propensity of any particular individual to survive, we find that the survivors get the propensity 1 and the ones that do not survive get propensity to survive of 0. This has the clear consequence that we cannot use the notion of propensity to distinguish between the survivors and the fittest conceptually by defining the fittest to be the ones with the greatest probability of survival. This shows that propensity does not always meet the requirement we wanted of the fittest, that is, the fittest may not be the ones that survive. In the deterministic case, we do not get the prospect that the fittest may not survive. In the deterministic case, the ones with the greatest propensity to survive simply will survive; they had a propensity to survive of 1, and those that do not survive will have had a propensity to survive of 0. This leaves no grist for the mill of natural selection. This seems to be an utterly compelling reason to look elsewhere for an account of fitness, particularly because at the time the Darwinian theory was born, the prospects for a deterministic physics seemed rather good. In that case, this sort of argument against Darwin should have looked devastating. However, it did not and still does not, so there must be another account of fitness to be had.

There are always possibilities of developing hybrid theories of probability; one is the theory of long-run propensities. After all, long-run frequency accounts simply relate to the presence or absence of conditions in comparison classes. One of those conditions can itself be propensities.

REJECTING THE IDEA OF FITNESS AS A PROBABILITY

We started off thinking that there is something right about the idea that we could define fitness in probabilistic terms. In particular, it seemed to meet the requirement that a variety might be the fittest and yet not be the ones that survived. In that case we said, they could have been the ones with the highest probability of surviving and yet still not survive, just as a die with five black faces and one white might come up white, although that was less likely than black. What we have seen, however, has dashed those hopes. Forced to consider what we meant by "likely" in saying that the fittest are the ones with the greatest likelihood of survival, we looked at each of the prevailing understandings and found that none of them answered to our theoretical needs. Probability just will not play the role we might have hoped it would play. Some have thought that Darwin's theory is best understood as involving the propensity account that turns on the fundamental physical level being indeterministic. Some have gone as far as to say that were underlying physics to be deterministic all bets are off and Darwin's theory falls to the side as useless. In the next section, I shall show why this is mistaken. Darwin's theory is not mortgaged in that way to fundamental physics. It depends on the world having a causal structure but does not depend on the nature of the underlying causal structure. It floats free of the underlying details, to at least some degree.

Bayesianism Does Not Save the Day

Bayesianism is a collection of methods for determining probabilities in accordance with the discovery of an eighteenth century mathematician, Thomas Bayes (1702–1761; Bayes and Price [1763]), of a fundamental theorem of chances now called "Bayes's rule":

$$P(A/B) = P(B/A)P(A)/P(B)$$

Provided P(B) is not zero.

Although it is sometimes described as an account of probability, it really is no such thing. Probabilities are assigned in accordance with the rule, but the nature of what is assigned is not determined by the rule. This is obvious in hindsight because there are subjective Bayesians and objective Bayesians who both follow Bayesian principles.

The subjective Bayesians are subjectivists who determine their confidences on the basis of Bayesian updating, suggesting that that there will be convergence of opinion in the long run even if we start with random prior assignments. This might seem attractive to subjectivists, but the long run is a long way away, and it is just not true that convergence is inevitable in the long run. In any case, once again there is the issue of the probability being in the wrong place. The dynamics of the world are somehow to be explained by confidence levels. More likely, the explanation of the dynamics will be some objective feature that is also supposed to explain why the confidence levels are where they are. Thus, the confidence levels are not playing the explanatory role.

Objective Bayesians follow the Bayesian rule and methods constructed in accordance with the rule and understand the method as discovering the objective chances, again not as an account of that chances are. Chances are still liable to problems discussed earlier, in particular the issue of what happens under determinism. In determinist systems, the objective chances of particular events are all driven to zero and one. Defining fitness for types of events by taking their means fails as an account of fitness. The reason is something we already discussed. The problem case to consider, as we said at the outset, is the one in which the fittest are the ones who by chance do not survive. If they do not survive, then the mean chance of survival of the fittest will still be lower than that of the less fit variety. Thus, mean chances will not play the role of defining fitness.

TENDENCY TO SURVIVE—FITNESS AS A DISPOSITION

The passages in which Darwin discusses fitness are often couched in the language of chance, the language of probability. Almost as often, he used another term, "tendency," which invites an altogether different sort of understanding. Tendencies are dispositions and dispositions are modal properties—ways things could, would, or should be under certain circumstances. Taking our lead from Darwin's characterization of fitness as a tendency, we can focus on the modal character of a tendency and try to understand fitness as a dispositional property. In particular, we can understand fitness as the disposition to survive relative to an environment.

What is a disposition? Dispositions are a kind of property. They are properties that relate to ways things could or would be under certain circumstances. This is captured in the very way we talk about dispositions by adverting to how the thing would be; something has the property of fragility when it has the disposition (or tendency) to break. Something has the property of solubility in water when it has the disposition to dissolve in water. It has that property regardless of whether it has ever been or will ever be near water. Dispositions then are properties things have, which are about how they would or could be. There is a difference between having the disposition and exercising it. A glass might be fragile and yet never break. A lump of sugar might be soluble—have the disposition to dissolve in water—although it is never placed in water and never dissolves. This is an attractive aspect of dispositions for the theory of natural selection. We needed a property that could be possessed by the fittest even if they do not end up being the survivors. Having the tendency to survive is perfectly consistent with not surviving on some occasions.

Philosophers have spent a lot of time discussing the nature of these dispositions just because they seem to be odd properties: Dispositions have something iffy about them. Dispositions seem to be properties that cannot be discerned directly but only by their being exercised, or perhaps by the identification of their underlying ground. They seem to point to something unreal, their merely possible exercising. Similarly, philosophers distinguish between the disposition possessed by an object and the ground of the disposition in the object. The ground of a disposition is the combination of properties of an object, which give it that disposition. If we consider the disposition, which is the fragility of a glass, we might say that the fragility is grounded in physical properties of the material it is made of, the particular crystalline structure that makes the glass vulnerable to breakage. Some philosophers have wanted to identify the disposition with its ground. In any case, there is an explanation of the way the object has the disposition in terms of the properties that ground the disposition. This too is an important insight into fitness and the way the other properties of an organism figure in grounding its fitness. This makes sense of our biological inquiry: we are trying to make sense of the way the different properties of an organism do affect the overall tendency of the organism to survive.

The disposition is stronger in some variants than in others. How much sense can we make of the notion of the strength of dispositions? Consider a disposition such as fragility, or poisonousness. We do ordinarily understand the notion of one object being more fragile or more poisonous than another. What do we mean by the notion of a stronger disposition?

Part, but only part, of what we mean by saying that an object, c, has a stronger disposition to F than another object, d, is that across a range of circumstances, c will have F in more of the circumstances. In measuring toxicity, we compare amounts of the substance, which will elicit a fatal result, the remarkable fact that 1/14,000th of an ounce of the venom of an Eastern Brown snake is a lethal dose for an adult human, for example, or again the number of mice that a given measure of toxic substance will kill. The inland taipan (*Oxyuranus microlepidotus*) holds the record for the number of persons it could kill with one bite worth of venom. The volume of venom discharged in one bite (110 mg) is enough to kill roughly 100 adult humans or 250,000 mice. It has been determined to be 10 times as venomous as the Mojave rattlesnake (*Crotalus scutulatus*) or 50 times more venomous than the common cobra (*Naja naja*). It has an LD50 of 0.03 mg/kg for mice, a standard measure of toxicity of venom. Even in clear cases such as the degree to which venom is a danger, the standard measures may be misleading. Venom does not react with all mammals the way it does with mice. Some snakes, for example, are evolved to hunt mice in particular. Some snakes are hunters of other snakes. Is a particularly efficient venom that kills mice an accurate indication of danger to humans or not? Quibbles like this aside, we can see that we do use the idea of stronger and weaker degrees of the same disposition, moreover that using it does affect the way we comport ourselves about the world. We know we have to step more carefully around people who are more tempestuous or explosive. We know that fragile wine glasses require different treatment than do beer steins.

Any way of explicating the idea of stronger and weaker dispositions is going to be relativized to circumstances. However, this particular way of spelling out the intuitive notion of a stronger disposition is not going to stand too much rigorous scrutiny. For one thing, it is very unlikely that we will be able to really depend on the notion of "more of the circumstances." Different ways of carving up the space of circumstances will lead to different numbers of circumstances. This really was the source of the Bertrand paradox we discussed in relation to the logical approach to probability: different ways of presenting the same situation lead to different assessments.

When we are dealing with a comparison of possibly infinite classes of circumstances, straightforward cardinality is unlikely to sort out the intuitive notion of "more." Intuitively, we would like to say that there are more integers than there are prime numbers, but in fact it is correct to say that there are no more integers than there are prime numbers. This is so although every prime number is an integer and lots of integers are not prime numbers: the set of primes and the set of integers have the same cardinality. In the case of the prime numbers, we can salvage the intuitive sense of there being more integers than primes by noting that the primes are a proper subset of the integers: every prime is an integer, but some integers are not primes. The set of primes is therefore properly contained within the set of integers. However, it is implausible that the range of situations in which the object with the stronger disposition exercises that disposition properly contains the set of the situations in which the object with the weaker disposition exercises its disposition. For there are always going to be circumstances in which a less fragile object breaks and a more fragile object does not. Perhaps one such scenario is one in which a very fragile object, because it is very fragile, attracts the attention of someone nearby, who manages to catch it when it falls. The less fragile object, because it is less fragile, leads to a careless attitude that leads to its breaking. This is not surprising and not going to show that the object that breaks is more fragile than the other object that did not. There are always going to be stories like this, ensuring that the superset relationship is not what is meant by the strength of the disposition. It would simply be dogmatic to assert that it never arises that the less fragile object breaks and the more fragile does not.

Another approach is to discuss what occurs in a typical or normal situation. This approach is promising, it seems to me, but not as an analysis because it seems to ride on antecedent commitment about the strength of dispositions rather than being an independent way of determining those. The notion of a typical or normal situation has the sort of vagueness in it that undermines any straightforward analysis of the notion of a stronger disposition.

Perhaps a doubly modal notion is needed here. For instance, the stronger the disposition, the easier it is for the disposition to be realized relative to a context. However, what is meant by this notion of easier realization? Perhaps something like this is envisaged: we imagine a context that we hold fixed except for the trigger of the disposition. Then a stronger disposition is possessed by the object that elicits the realization of the disposition with a wider range of triggers/weaker range of triggers. The use of fragility or poisonousness as examples might lead to the thought that the strength of the disposition can be assessed by the amount of shearing pressure before breakage in the one case and amount of substance ingested before damage is done in the other.

This is suggestive but probably rests on our sense of strength of dispositions again rather than spelling out what we mean by strength of the disposition.

Thus, developing an analytical theory about strength of dispositions is not easy. Luckily, we do not need to do that to justify the claim that we readily understand and use the idea of strengths of dispositions. We do, and that is all we need here to make sense of the idea that fitness is a disposition that comes in different strengths.

One interpretation that someone might feel tempted to give of the possibility of the less fragile object breaking in a situation in which the more fragile object does not is to trade on the notion of probability. The idea might be that in that circumstance, the more fragile object has a higher probability of breakage than the less fragile object. However, we started out trying to understand this very notion of probability. We have seen that it is implausible that it can be made sense of by any of the accounts of probability we have looked at. Certainly, if the world were indeterministic (and our best physics tells us it is), we could use the notion of objective chance. However, it is most important to see that it is a mistake to hold that anyone who uses the notion of one object being more fragile than another is committed to indeterminism at the level of physics. A commitment to dispositions is neutral on the matter of physical indeterminism. Objects have dispositions, and so the relevant counterfactuals are true regardless of the issue of determinism and indeterminism.

What if the terms we must use to define the notion of fitness are inextricably modal? There is a historical distrust of modal notions. But such notions are fundamental to our understanding of the world and although they are not exhausted by empirical tests are nevertheless subject to empirical inquiry. This is not a recipe for despair. For example, that is not to say that we do not have empirical reasons for the ascription of these modal properties. We can and do investigate the properties which ground dispositions to break, to harm if ingested, to kill, to survive, and so on. We can investigate the properties empirically. However this does not show that these properties have an empirical nature. To make that leap of thought is to confuse the evidence we have with what the evidence is evidence for. This is an important distinction. The dissolving of the grains in water is good evidence that the grains were soluble, but that dissolving is not the solubility. It is the manifestation of the disposition, not the disposition. Moreover, objects can have that disposition to dissolve while they never manifest that disposition. The disposition is possessed and explains the dissolving. However, the disposition does not float free. It is grounded in the properties of the grains. We seek explanations of disposition in terms of other properties, and that is an important part of our trying to understand how objects behave. When it comes to dispositions, such as the disposition to poison, some of the underlying properties that ground that disposition can themselves be other dispositions. For example, solubility is one disposition that is part of the grounds for being poisonous for some substances.

Fitness can be characterized as the tendency to survive. The tendency to survive will depend on the way features of the organism ground that tendency. That is, the degree of adaptation is degree to which the features possessed tend to assist persistence of those features. This notion of adaptation is independent of the notion of natural selection. We do not define the degree of adaptation in terms of the historical origin of the features with which we are concerned. It is for this reason that we can use the notions of adaptation and fitness in the characterization of natural selection.

So far, we have seen that it is a mistake to define the fittest as those that survive. It is also a mistake to define the fittest in terms of actual survival numbers, including the common accounts of fitness in terms of those that differentially survive. Such an account fails to explain the notion of fitness adequately. This much is not controversial. What is controversial is the reason for the inadequacy of such an account of fitness. I have tried to suggest the reason: that to define fitness or adaptation in terms of actual survival would strip the theory of natural selection of its explanatory power if it were adopted. We tried the idea that the fittest are those with greatest probability of survival, but that foundered when we looked closely at the different notions of probability that might underpin that claim. We have seen some of changes we might introduce to deal with this problem and why the dispositional account seems the most attractive candidate. The second premise also needs attention.

DEFINING NATURAL SELECTION

Premise 2 of the simple circularity argument was that natural selection defined to be the survival of the fittest (Figure 3.1). This premise also played a key role in the circularity argument, and this premise in fact has largely avoided scrutiny in the literature. The phrase "the survival of the fittest" is both simple and seductive. It is very commonly treated as meaning the same thing as "natural selection," but not without reason. After all, Darwin himself seemingly legitimated that identification when he called the pivotal fourth chapter of *On the Origin of Species*, "Natural Selection, or the Survival of the Fittest." The phrase "the survival of the fittest," however, is not originally his and does not occur in the first edition of *On the Origin of Species*. The phrase was coined by Herbert Spencer (1864, 444–445), who says, "This survival of the fittest, implies multiplication of the fittest … This survival of the fittest, which I have here sought to express in mechanical terms, is that which Mr. Darwin has called 'natural selection,' or the preservation of favoured races in the struggle for

FIGURE 3.1   Herbert Spencer.

life." Spencer then clearly thought he was coining a phrase for the concept Darwin had already fashioned (Figure 3.2). Darwin seized on the phrase and used it for the first time in the fifth edition of *On the Origin of Species* published by John Murray in 1869, or ten years after its first edition. Darwin acknowledged that he got the phrase from Herbert Spencer and used it in all later editions of *On the Origin of Species*. Interestingly, he addressed the two phrases in the fifth edition:

> I have called this principle, by which each slight variation, if useful, is preserved, by the term Natural Selection, in order to mark its relation to man's power of selection. But the expression often used by Mr. Herbert Spencer of the Survival of the Fittest is more accurate, and is sometimes equally convenient. (Darwin, 1869, 72)

> vive whose functions happen to be most nearly in equilibrium with the modified aggregate of external forces. ·
>
> But this survival of the fittest, implies multiplication of the fittest. Out of the fittest thus multiplied, there will, as before, be an overthrowing of the moving equilibrium wherever it presents the least opposing force to the new incident force. And by the continual destruction of the individuals that are the least capable of maintaining their equilibria in presence of this new incident force, there must eventually be arrived at an altered type completely in equilibrium with the altered conditions.

FIGURE 3.2   First occurrence of the phrase "survival of the fittest" in Spencer's *Principles of Biology*, Vol. 1.

It may suggest an image of evolution as a vigorous and bloody fight for survival among varieties with the fittest left standing triumphantly over the carcasses of the less fit. This image certainly fits the caricature so often offered of nature as a war of all against all, of evolution as a revolution that eats its own children, or better, grinds on over their broken bodies. Moreover, these images are redolent with the then pervasive nineteenth century smugness that the status quo represents not just the outcome of progress but even the highest form achievable. Those images did not lead Darwin to choose the phrase and in fact to prefer it to the phrase he coined "natural selection." In fact, it was quite the opposite. He showed that he was uncomfortable with the images carried by his own phrase, in particular, with the false suggestion it carried that nature was a force that showed foresight and even prudence in making choices among the varieties of life parading before it. Darwin spent much time trying to distance his concept of natural selection from this image of nature as judge and executioner. He seems to have preferred the Spencerian phrase "survival of the fittest" for the absence of such a suggestion. That phrase carries no suggestion that there is an intelligence that selects which survive and which do not. Rather more accurately, the suggestion is that evolution by dint of the survival of the fittest varieties is a process that does not require any external intervention. The triumphalist Victorian imagery often associated with his theory has also been used to play down the originality of Darwin's achievements in developing the theory, to associate Darwin's theory with the justification of nineteenth century imperialism, and to link Darwin's theory with the negative social philosophy called "Social Darwinism."*

These are, one and all, mistakes. In fact, it is a mark of Darwin's extraordinary intellectual achievement that he did not succumb to the progressivist conceptions of evolution. It would have been all too easy to fall in with the prevailing and almost overwhelming conception of progress. Darwin's own work was marked by a tendency to see the evolutionary changes as occurring in the direction of both increasing complexity and increasing simplicity. His most weighty work, in terms only of pages, was his landmark study of barnacles. Perhaps it was the study of these enigmatic creatures whose development is marked by a move from a motile larval form to a sedentary adult form that inoculated Darwin against the prevailing prejudice in favor of progress. In his study of barnacles, he saw that sometimes selective pressures favor varieties that are simpler and less specialized than their predecessors.

What we will turn to now is the undermining of the central role the phrase "the survival of the fittest" has played in characterizations of the theory of natural selection. However attractive the dramatic images associated with that phrase may have been to the bourgeois of the mid–nineteenth century, Darwin's use of the phrase "the survival of the fittest" was an unfortunate distraction from the explanatory structure of his theory. Darwin used it as he saw the phrase as an answer to the unhappiness he felt with the suggestions of intelligence and intention carried by his own phrase "natural selection." However, for all its pithy elegance and attraction, there is something

---

* See in particular the work of Gertrude Himmelfarb (1968).

powerfully misleading about the phrase "the survival of the fittest," something that continues to mislead.

The premise of the circularity argument that defined natural selection as the survival of the fittest talked all too simplistically as if survival is what matters. This obscures certain features of the theory of natural selection. Fitness is, as we have seen, a measure of the tendency of survival or persistence. Even in Darwin's time, there was no suggestion that the fittest varieties survived and the less fit did not. Rather the picture was that relatively more the fitter varieties survived to reproduce, and so more of them got to be present in the next generation. Thus, plenty of less fit varieties survived; what marked off the fittest is that they disproportionally survived, or as we say today, they "differentially survived." That phrase understood that way is fine and captures what many biologists still conceive as the model of natural selection. They should not. I want to show you that natural selection should not be defined as the survival of the fittest, nor is the modern alternative, the differential survival of the fittest, in any better stead. I also want to show that Darwin himself did not seem to treat natural selection as equivalent to the Spencerian phrase in his discussion of actual cases.

An event that is the greater and disproportionate survival of those with greater tendencies of survival than of those with lesser tendencies is not as such sufficient for natural selection. This is because this distribution may not be due to the features that ground the tendencies to differential survival in the right way. Meteors, which by chance wipe out the less fit, do not thereby bring about an event of natural selection. I emphasize the important phrase "by chance." If some organisms evolved particularly hard heads under a constant shower of these meteors, we could indeed have a case of natural selection. Or indeed, if some sonar detector evolved under other selective pressure makes the organism cringe in a cave when meteors approach, then again we can have natural selection. This is to be contrasted with mere chance, the fortuitous happenstance of being in the right place at the right time.

We have seen that the fittest individuals may survive for many reasons, not all of which may be explained by those characteristics that make them the fittest. This means that not all instances of the survival of the fittest are instances of natural selection. A purely fortuitous event, even if it results in the differential survival of the fittest, is not natural selection for the simple reason that it made no difference that it had the properties that make it the fittest. The features that made them the fittest did not explain their differential survival, so adverting to the features would not explain why these survived and the alternatives did not. For that reason then, we have to reject the idea suggested at various points by Darwin himself, that is, "survival of the fittest" is simply a better form of words for what he had been calling "natural selection." The difference is in one being the mere survival of the fittest for whatever reason and the other being the survival of the fittest that is explained by their having the properties that make them fitter.

The distinction between natural selection and survival of the fittest may seem rather recherché, but distinguishing between the two is required by the explanatory structure of the theory of natural selection. Not every instance of survival of the fittest will be an example of natural selection. Moreover, there is some textual support in Darwin's work for something very close to this distinction.

It may be well here to remark that with all beings there must be much fortuitous destruction, which can have little or no influence on the course of natural selection. For instance a vast number of eggs or seeds are annually devoured, and these could be modified through natural selection only if they varied in some manner which protected them from their enemies. Yet many of these eggs or seeds would perhaps, if not destroyed, have yielded individuals better adapted to their conditions of life than any of these which happened to survive.* (Darwin, 1872, 68)

In the final sentence, Darwin indicated that the fittest or best adapted individuals may have died and may not be the individuals that happened to survive. He was clearly not equating surviving with being the fittest. This passage also clearly indicates that Darwin did not equate natural selection with the survival of the fittest. As he tells us, some of the individuals fortuitously destroyed may indeed have been the fittest, and equally, we may surmise that some may have been the least fit, but this can have no relevance to natural selection unless they "could be modified through natural selection only if they varied in some manner which protected them from their enemies." The fact that the individual that would have developed from the egg was in fact less fit does not make such an instance an example of natural selection. Thus, Darwin cannot be criticized in an unqualified manner for adopting Herbert Spencer's phrase. Although his adoption of the phrase did mislead some who have read and thought about natural selection, there were clear articulations of his views, which ran counter to the phrase. This passage and many like it, seemingly disregarded by many commentators, also make clear that in *On the Origin of Species*, the essential elements and many of the details of a properly articulated account of a theory of evolution by natural selection are found and can still be endorsed today.

Natural selection is not to be defined as the survival of the fittest because there are cases of survival of the fittest that are not cases of natural selection. There is more going on in natural selection than the mere survival of the fittest. What more is going on? Focusing on the way we explain changes by natural selection tells us what more is going on. When we use the mechanism of natural selection in explaining evolutionary change, we are pointing to the causal relevance of the features we think make the variety fitter. For it to be an example of natural selection, the features that make the variety fitter have to explain the difference in survival. That is enough to show that, in the example Darwin considers on the destruction of eggs and seeds, if the less fit individuals the ones that were differentially destroyed, it would not be an example of natural selection unless there were features that the fitter individuals possessed that explained their differential survival. In Darwin's example, there were no such features; it was merely fortuitous. In that case, although that destruction of the seeds or eggs was the differential survival of the fittest, it was not an example of natural selection.

Thus, in that case, we are now in a position to more accurately characterize natural selection. Natural selection is the differential survival of the fittest caused by the features that make them fitter. This explains why discovering whether something

---

* This passage does not appear in earlier editions and is in the final edition of *On the Origin of Species* published in 1876 as well.

arose by natural selection is so difficult. To be confident that something arose by natural selection, we need to know that it was fitter, which is difficult enough, but we also need to have reasons to believe that the causal history fits this sort of account, that the feature was explanatory of a differential rate of survival.

Fitness is a key concept for Darwin. It is what explains the survival of certain types, all else being equal. The account we have reached is that fitness is a disposition to survive. That disposition comes in different degrees. Some varieties have a stronger disposition to survive than other varieties. Moreover, that disposition is grounded in the properties of the varieties. One of the key things biologists investigate is the way the different properties are involved in affecting that disposition to survive. One of the tricky aspects of dispositions is that they are hard to discern. If they are hard to discern, how are we to measure them? This becomes a pressing problem. If we are right about what fitness is, the claim that variety A is fitter than variety B means that variety A has a stronger disposition to survive than variety B. That claim is consistent with variety B surviving. As we have seen, it is clearly possible for the fittest not to be the survivors. The problem immediately poses itself: How can we measure fitness? The common answer is to operationalize the concept of fitness. Treat it as not what explains survival but survival itself, or rather something we can more easily measure. Dispositions are tricky, in the same way causation is tricky, and David Hume showed just how tricky the notion of causation was.

CONCLUSION

Although the idea that fitness is reproductive success is very widespread, we can see that it is a mistake. That definition confuses fitness with what it tends to produce. Such an identification confuses success with the properties an organism has that lead to that success. Fitness is a relationship between an organism and its environment, which tends to help the organism survive and often leads to survival. The notion of a tendency is not what happens on average; it is a dispositional notion. Moreover, we have seen that natural selection is not reproductive success, and it is not the survival of the fittest. Although the Darwinian credentials of the equation of the survival of the fittest with natural selection may have seemed impeccable, we have seen that this equation must be rejected, both as not reflecting Darwin's considered view and as in fact false. Natural selection entails but is not entailed by the survival of the fittest. Natural selection is the survival of the fittest because of the features that make them the fittest. Thus, natural selection is not, after all, the survival of the fittest.

REFERENCES

Bayes, T. and R. Price. 1763. An essay towards solving a problem in the doctrine of chance. By the late Rev. Mr. Bayes, Communicated by Mr. Price in a letter to John Canton, A.M.F.R.S. *Philosophical Transactions of the Royal Society of London* 53: 370–418.

Benson, R. B. J., T. H. Rich, P. Vickers Rich, and M. Hall. 2012. Theropod fauna from Southern Australia indicates high polar diversity and climate-driven dinosaur provinciality. *PLoS One* 7(5): e37122. doi:10.1371/journal.pone.0037122.

Bigelow, J. and R. Pargetter. 1987. Functions. *Journal of Philosophy* 84: 181–196.

Carnap, R. 1950. *Logical Foundations of Probability*. Chicago: University of Chicago Press.

Carnap, R. 1952. *The Continuum of Inductive Methods*. Chicago: University of Chicago Press.

Carnap, R. 1963. Replies and systematic expositions. In *The Philosophy of Rudolf Carnap*, edited by P. A. Schilpp. La Salle, IL: Open Court.

Darwin, C. 1859. *On the Origin of Species*. 1st ed. London: John Murray; 2nd ed., 1860; 3rd ed., 1861; 4th ed., 1866; 5th ed., 1869; 6th ed., 1872; 6th ed., with additions and corrections, 1876.

Dennett, D. 1995. *Darwin's Dangerous Idea*. New York: Simon and Schuster.

Gillies, D. 2000a. Varieties of propensity. *British Journal for the Philosophy of Science* 51: 807–835.

Gillies, D. 2000b. *Philosophical Theories of Probability*. London: Routledge.

Hájek, A. 2012. Interpretations of probability. In *Stanford Encyclopedia of Philosophy*. Winter 2012 ed., edited by E. N. Zalta. http://plato.stanford.edu/archives/win2012/entries/probability-interpret/, accessed on March 12, 2014.

Himmelfarb, G. 1968. *Darwin and the Darwinian Revolution*. 2nd ed. London: Chatto and Windus.

Lewontin, R. 1974. *The Analysis of Variance and the Analysis of Causes*. Reprinted as Chapter 4 of Levins and Lewontin (1985). Cambridge: Harvard University Press.

Lewontin, R. 1978. Adaptation. *Scientific American* 239: 156–169.

Lewontin, R. 1985. Adaptation. In *The Dialectical Biologist*, edited by R. Lewontin and R. Levins. Cambridge: Harvard University Press.

Maynard Smith, J. 1976. Evolution and the theory of games. *American Scientist* 64: 41–45.

Maynard Smith, J. 1979. Game theory and the evolution of behaviour. *Proceedings of the Royal Society of London, Series B* 205: 475–488.

Maynard Smith, J. 1983a. *Evolution and the Theory of Games*. Cambridge: Cambridge University Press.

Maynard Smith, J. 1983b. Adaptation and satisficing. *Behavioral and Brain Sciences* 6: 370–371.

Millikan, R. 1984. *Language, Thought and Other Biological Categories*. Cambridge, MA: Bradford/MIT Press.

Millikan, R. 1989a. In defence of proper functions. *Philosophy of Science* 56: 288–302.

Millikan, R. 1989b. Biosemantics. *Journal of Philosophy* 86: 281–297.

Mills, S. and J. Beatty. 1979. The propensity interpretation of fitness. *Philosophy of Science* 46: 263–288.

Neander, K. 1991a. Functions as selected effects: The conceptual analyst's defence. *Philosophy of Science* 58: 168–184.

Neander, K. 1991b. The teleological notion of function. *Australasian Journal of Philosophy* 69(4): 454–468.

Paley, W. 1836. *Natural Theology*. Vol. 1. London: Charles Knight.

Peters, R. H. 1976. Tautology in evolution and ecology. *American Naturalist* 110: 1–12.

Popper, K. R. 1957. The propensity interpretation of the calculus of probability and the quantum theory. In *The Colston Papers*, edited by S. Körner. Vol. 9, pp. 65–70, London: Butterworths.

Popper, K. R. 1959. The propensity interpretation of probability. *British Journal of the Philosophy of Science* 10: 25–42.

Ramsey, F. P. 1931. Truth and probability. In *Foundations of Mathematics and other Essays*, edited by R. B. Braithwaite. London: Kegan, Paul, Trench, Trubner, & Co. pp. 156–198; Reprinted in *Philosophical Papers*, D. H. Mellor (ed.). Cambridge: Cambridge University Press, 1990.

Sober, E. 1984. The *Nature of Selection*. University of Chicago Press: Chicago.

Spencer, H. 1864. *The Principles of Biology*. Vol. 1. London: Williams and Norgate.

Stoppard, T. 1967. *Rosencrantz and Guildenstern Are Dead*. New York: Grove Press.
van Fraassen, B. 1989. *Laws and Symmetry*. Oxford: Clarendon Press.
Wallace, A. R. 1914. *The World of Life; A Manifestation of Creative Power, Directive Mind and Ultimate Purpose*. London: Chapman and Hall.
Woodward, H. N., T. H. Rich, A. Chinsamy, and P. Vickers-Rich. 2011. Growth dynamics of Australia's polar dinosaurs. *PLoS One* 6(8): e23339. doi:10.1371/journal.pone.0023339.
Wright, L. 1973. Functions. *Philosophical Review* 82: 139–168.

4 Darwin's Key Argument for Evolution by Natural Selection

THE ARGUMENT DARWIN USES

In *On the Origin of Species*, Darwin presented in various forms, a key argument for evolution by natural selection. One example of the argument is the following from the first edition of *The Origin* (1859, p. 5)*:

> As many more individuals of each species are born than can possibly survive; and as, consequently, there is a frequently recurring struggle for existence, it follows that any being, if it vary however slightly in any manner profitable to itself, under the complex and sometimes varying conditions of life, will have a better chance of surviving, and thus be *naturally selected*. From the strong principle of inheritance, any selected variety will tend to propagate its new and modified form.

This argument is very much worth pulling apart. Seeing how it works will show us a lot about the structure of the theory of natural selection as Darwin understood it.

The argument rests on a number of premises. The first is excess, more individuals are born than can possibly survive. The second, which he claims follows from the first, is competition, the struggle for existence. These two premises are interesting in their own light and some interesting questions arise that are not easy to answer about the role of natural selection in contexts in which there is no competition. The Russian anarchist and important ecological thinker Peter Kropotkin (1902) suggested that in the tundra, for many organisms, it is not a competition for limited resources that shapes the ecosystem but rather the ability to expand quickly into unoccupied environments as conditions change from harsh winters to less harsh spring and summer. Correct or not, this is an interesting challenge to a competition-focused approach to evolution. Moreover, there are ways in which changes arise in populations that seem to be unrelated to competition. The genetic make-up of a population can change simply because, in the absence of competition, the types of individuals that produce the most offspring will come to increase their proportion within the population, other things being equal. We can put off discussing this sort of scenario right now but it ought to be clear that this sort of change does not seem to be related at all to the degree of adaptation organisms have to their environment.

The third premise in Darwin's key argument is, in many ways, the most interesting. He says "if some of the variations among individuals can be beneficial in the

---

* These are all (and much more material) very usefully available online in searchable format at darwin -online.org.uk/contents.html.

struggle for existence, those individuals have a better chance of surviving." This is one of the key occurrences of the term "chance" in Darwin's text and we have seen that making sense of this claim as being about some notion of probability is not possible. On the other hand, the term chance is closely aligned with the notion of a tendency. We've seen that we can make sense of fitness in terms of a tendency or disposition to survive. What is striking about this third premise is that Darwin does not assert that variations exist in a population or that variations that do exist differ in fitness. Rather, he makes a very cagey move. He says that *if* some variation can be beneficial in the struggle for existence, those will have a better chance of surviving, or as we might put it, have a stronger disposition to survive, or be fitter. Notice then this is an analytic claim. This is not empirical at all.

The fourth principle is that of heredity; variations present in parents tend to be present in offspring too.

This argument is quite fascinating and Darwin seems to have thought very highly of it because he presented it many times in *The Origin*. Although successfully, there are quibbles that need to be discussed but the important thing to note is just how persuasive and general the argument is. Moreover, by looking at the way the concepts of competition, selection, and inheritance work in this argument, we get a better picture of what these notions need to be for Darwin's theory to work. This is important because as our knowledge has grown, our understanding of the mechanisms underpinning, for example, biological inheritance has become immeasurably richer. However, sometimes the more detailed understanding gets in the way of the more general notion Darwin himself depended on in this argument. In effect, what is happening is that we mistake the mechanism for the function. To highlight just how the biological nature of the theory of natural selection enters into the story—and in so doing, what Darwin's specifically biological assumptions are—it is useful to contrast Darwin's account with a more general "proto-theory of natural selection," which is no more than a theory of persistence.

Darwin's argument then is as follows:

1. More individuals are born than survive to reproduce.
   Therefore,
2. There is a struggle for existence.
   Therefore,
3. If some variation were advantageous in the struggle to survive to reproduce, it will have a greater tendency to survive in the struggle for existence.
4. If it did survive, then it would have been an instance of natural selection.
5. If that advantageous variation were heritable, its offspring would tend to have that variation and so also have the advantage.

Premise 1 is an empirical claim. We can measure the number of individuals born and the number of individuals that survive to reproduce. There is nothing necessary about this claim but empirically it bears out. This overwhelmingly seems to be the case. Darwin here is using a perfectly empirical claim about the world.

Let's look at the inference from premises 1 to 2. There are situations in which premise 1 is not true but these are odd cases. Most often, not all of a generation survive to reproduce, so more are born than survive to reproduce. Thinking of it that way, premise 2 seems to be just another way of saying that there are more individuals born than survive to reproduce. If that is so, then the inference from premises 1 to 2 is assuredly valid.

What about premise 3? It too has a ring of obviousness about it, especially if we were to spell out what "advantageous" means in terms that make a variation have a greater tendency of survival as we should. So understood premise 3 is an analytic truth. It is spelling out the connection between advantage and the tendency to survive.

Thesis 4 claims that if an advantageous variation were to survive, then it would have been naturally selected. This, it seems, is the weakest part of the argument. What Darwin is saying here is not straightforwardly right. After all, might the advantageously bestowed not survive simply by chance? Would that be an example of natural selection? It seems not. We have seen reasons to think that Darwin himself accepted that there was much fortuitous surviving that was unrelated to fitness. So what more would he need? There are two ways we can understand what Darwin has said. The first is that the fittest merely surviving is enough for it to be an example of natural selection. The second is that he has some other background assumption about the usual run of things, perhaps that in the long run, the ones with the greatest tendency to survive will because of that survive. These are quite different although it might look like a subtle difference at first.

Consider the first interpretation of what is going on. If it were true that any instance of the fittest surviving is an example of natural selection, then we have to relinquish the idea that adverting to natural selection is an explanation of the different rates of survival. The key thing here is that because Darwin has told us that surviving can be fortuitous and so unrelated to fitness, mere survival is not itself enough to ensure that it was the fitness that explained the survival. Therefore, were this interpretation to stand once again, we would be in a position where saying that the fittest survived and saying the fittest survived because of natural selection, is saying the same thing twice over but the second time with the pretense of an explanation. If adverting to natural selection is adverting to an explanation for an evolutionary change then that interpretation cannot stand.

This can also be taken as a necessary truth. If after all, the definition of natural selection is the survival of the fittest, then clearly if the fittest individuals survive, then that is an instance of natural selection. In fact, this is a point at which I claim there is a weakness in Darwin's argument as it is presented. Later on, he wants to argue that it is because of the advantage conferred by the variation that that variation survived. He cannot both claim that it is the mere survival of the most fit variation that makes it an instance of natural selection and also say that evolution by natural selection occurs because of the advantage conferred to the fittest by the variation. These two claims pull the theory in two different directions that are still in tension in biological thinking today.

On the one side, by making fitness and natural selection measureable, we make the theory more clearly applicable to the world, and that sounds like a serious advantage. However, that approach does so by stripping the theory of any explanatory power. As we saw in the previous chapters, there is a cost of defining fitness and natural selection in a manner that is more empirically accessible. We no longer are in a position to say that being fittest explains the survival. After all, in this way of doing things, surviving is what makes that variety the fittest. This account as we have seen is inconsistent with the idea that the fittest are in any instance not the survivors. That has consequences for the notion of natural selection that we saw were unpalatable. Darwin himself was clear about his commitments and would not have been happy with the claim that it suffices to show that an instance of evolution is an example of natural selection showing that the most advantaged individuals survived. It certainly counts as evidence that it was an example of natural selection but it does not suffice to show that it was an example.

The problems with the first interpretation lead us to the second interpretation. Darwin must have had in mind the idea that, in the main, the fittest survive because the features that are advantageous do in fact help them to survive. In that case, and (again) in the main, the fittest survive because they have those advantageous variations. This needs an appeal to something like the law of large numbers: in the main, it is reasonable to think that the fittest with their advantageous variations will survive predominately.

Thesis 5 emphasizes the role of heritability and this key notion is also an empirical matter. It is utterly empirical whether the varieties that predominately survive have offspring that are like themselves. In Chapter 7, we will see that this notion of heritability is something that is often confused with its underlying mechanisms. It is perfectly clear that the question whether offspring are more similar to their parents than to the population as a whole is a straightforwardly empirical question. It is also something Darwin drew attention to and was the first to see the theoretical importance of, and yet heritability is something people had long noticed. Animal breeders had depended on this realization for their animal husbandry. Gardeners moving from place to place would choose to plant seeds from varieties that did well in different environments. Again, this claim of the heritability of variation seems overwhelmingly correct.

What do we get when we put this argument together? It is an argument that uses two empirical claims:

1. There are more born than survive to reproduce.
   and
2. Variations are heritable.
   It uses one analytic claim, true by dint of definitions:
3. If some variation were advantageous in the struggle to survive to reproduce, it will have a stronger tendency to survive to reproduce in the struggle for existence.

Moreover, it appeals to the law of large numbers, in parentheses because Darwin does not explicitly use this premise but it seems to be required:

(It is reasonable to think that the fittest with their advantageous variations will survive predominately)

The conclusion he draws is that:

If that advantageous variation were heritable, its offspring would tend to have that variation and so also have the advantage.

With this argument, Darwin spells out how given some well attested empirical claims, one reasonable premise (whose status we will explore later) and a definition we can see that over time the fittest will tend to survive because of their advantageous features. This is why the theory looks obvious. Once Darwin has pointed out the premises, some of which have been hiding in plain sight, unacknowledged but obvious once pointed out, a powerful engine for organic change has been recognized. As Thomas Huxley (1901, 183) put it:

> The facts of variability, of the struggle for existence, of adaptation to conditions, were notorious enough; but none of us suspected that the road to the heart of the species problem lay through them, until Darwin and Wallace dispelled the darkness, and the beacon-fire of '*The Origin*' guided the benighted.

The facts Darwin used as his premises were not arcane or unknown; they were plain and obvious to all. As Huxley suggests, what Darwin did was to see how these obvious facts were the key to understanding the way evolution could take place.

A PROTO-THEORY OF NATURAL SELECTION

Darwin's full theory of natural selection is essentially a theory of persistence. We have just seen that the point of the theory is to show how the properties of different varieties can explain the differing tendencies of the varieties to persist. The theory of natural selection as presented by Darwin involves a number of assumptions that relate to the biological case for which it is constructed. There was his premise about the heritability of variation, for example. There was his premise about the excess of offspring relative to reproducing individuals. Some of these assumptions can be relaxed, yielding a more general theory that delivers enough of the same sort of consequence also to be regarded as a theory of persistence. This is a proto-theory of natural selection. Moreover, by developing a more general theory of persistence, some of the key features of the theory of natural selection can be highlighted. We can, for example, find analogues for the concepts of fitness and of natural selection in this theory of persistence that enable us to see more clearly how these figure in explanations.

Let me parody the brilliant argument Darwin developed for natural selection to give shape to this more general theory.

> Things differ with respect to their properties. Among the properties of things, it would be remarkable if there were not some properties of some things which are beneficial to their persistence over time and some properties of some other things which are injurious to their persistence over time. In that case, more of the former than the latter will tend to persist over time.

There is nothing contentious about this theory of persistence. It spells out an assumption, that some properties of things will help them persist and that some will be injurious to their persistence. Then, it concludes, on the basis of some principle such as the law of large numbers, that more of the things with properties beneficial to their persistence will persist than the things with properties injurious to their persistence. This much is fairly trivial and makes sense of C.S. Peirce's disparaging remarks that Darwin's theory is just the application of the law of large numbers to biology (Peirce, 1877, 7):

> The Darwinian controversy is, in large part, a question of logic. Mr. Darwin proposed to apply the statistical method to biology ... In like manner (to statistical reductions of gas thermodynamics) Darwin, while unable to say what the operation of variation and natural selection in any individual case will be, demonstrates that in the long run they will, or would adapt animals to their circumstances.

This indicates that Peirce's understanding of the structure of the theory of natural selection was rather limited. The work that is done by Darwin's and even the more general theory of persistence, as we have seen in this chapter, is not trivial. Let me demonstrate with some examples of how the more general theory might be deployed. First, an example from the domestic arena: drinking glasses. If we have a range of glasses and they are used for all sorts of drinking and are used equally frequently, then we would expect, in the long run, that those glasses that are less fragile would survive in greater numbers than those glasses that are fragile. Fragility is the tendency the object has of breaking. Of course, knowing this, we might expect that we will tend not to use the very fragile ones except on very special occasions, and certainly never let the children near them. Moreover, we could probably offer intelligent opinions about why some glasses tend to break more easily than others. The tall thin champagne glasses are readily broken: their height and narrowness means they are more liable to fall over. They are more often used standing up so they have further to fall, and so the story goes. When we explain the different rates of persistence, we do that by spelling out the sorts of properties the different types of glasses have that lead to their having a differential tendency to break. These are claims about what sorts of properties the glasses have and how those properties interact with their behavior to cause their breaking.

A second example comes from the world of publishing. Paper is made by breaking down plant material into cellulose fibers and reconstituting it into a flat form. Paper has been used for centuries and had been made of various plants, mulberry was a favorite. The long fibers of the mulberry, for example, made good quality paper that lasted a very long time. Until the middle of the nineteenth century, wood pulp was not used to make paper but at that point it was realized that wood pulp that had been broken down with acids could be used to make paper and that became a cheap and common source of paper. By the 1930s, however, people started noticing that books more than a few years old were becoming brittle and yellowed. Pages folded over would not crease; they simply snapped at the fold. Moreover, things got worse if the books had been exposed to sunlight. This was new. Books made of paper had not had this problem before. It turned out that the books so afflicted had been made from acidic wood pulp paper. These books had been printed on paper bleached white by the use of acids. Although much of the acid was washed out as a part of the production process, there nevertheless remained some acid in the paper. The moisture in the air interacting with the residual acid was enough to make the paper first turn the pages yellow, then brittle, and then turn into dust.

The persistence of these books or rather the lack of persistence of these books is explained by, among other things, the underlying properties of the materials out of which the books were constructed. Given two books produced in the same year, the one using paper produced by the acid bleaching process would have a lesser tendency to survive to any given year in the future. On the other hand, these acid-bleached books were cheaper to produce so they were produced in larger numbers. So now there is a trade-off, each particular book is less likely to survive if it was produced with acid-containing paper but on the other hand if enough copies of the book were produced, it is more likely that some or one of them survives to exist.* What is interesting is the way that we explain the persistence or lack of persistence of a book by using the features that the book has that either helps it or harms its tendency to persist. That is perfectly in keeping with the proto-theory of natural selection.

The explanatory role of this sort of theory of persistence is also well illustrated by the nature of the fossil record of very early Therian mammals, in particular from the Early Cretaceous. Much to my own disappointment, this fossil record consists largely of isolated teeth and small fragments of tooth bearing bone. Of the teeth which survived, molariform teeth are present in the fossil record with far greater frequency than are precanine teeth. When Bryan Patterson conducted his landmark study on the Trinity Formation from the Aptian period of the Early Cretaceous near Forestburg, Montague County, Texas, he found some 28 specimens to examine after years of close scrutiny of the sands of the Trinity formation. These were all molariform teeth. What happened? How is it that when the animals died more than 110 million years ago, we went from a whole animal to isolated molars?

When an animal dies and is going to end up as a fossil, it will typically be buried in mud or fine sand. In the Trinity formation, it was sand. At the time of death, of course, the animal was made up of soft and hard tissue. The soft tissue usually does not last long. Although there are rare and wonderful imprint fossils that record even the fine hair-like feathers on some Therapod fossils, typically soft tissue decays quickly and over time the harder skeletal material remains because it resists decay better. Notice that we can explain that difference in persistence between the hard and soft tissue by pointing to differences in the properties of the soft tissue and the harder tissues. When the soft tissue has decayed, the hard tissue, the skeletal material, remains. Of that skeletal material, we find that there are bones of different shapes. If the skeleton is laid down and undisturbed, we can sometimes find complete skeletons. However, if the ground is disturbed at all, the shape of the bones can be very important to whether the bones survive. For example, ankle and wrist bones are roughly spherical, at least when compared to long slender bones such as ribs and fine appendage bones. The shape of the ribs makes them more likely to break. Ankle bones, on the other hand, are more resistant to shearing pressures and so tend not to break. Thus, over time, we expect to see among the bones more of the bones shaped so as to resist breaking, and this is what we do tend to see. On the other hand, not

---

* I mean this to contain echoes, explanatory echoes, of the well-known difference in strategies taken by *K*- and *r*-selected species in ecology. *K*-selected species produce few offspring and put a lot of resources into them; *r*-selected species produce many offspring and put less resources into each offspring.

all hard tissue is equally hard: teeth are made of harder material than are bones. The material teeth are made from wears less and breaks less than bone does. Over time, we would expect that of the material that was left, more of the harder less fragile teeth would persist. And of the teeth? An example is afforded by the early Aptian (early Cretaceous) Therians studied by Bryan Patterson (1956). These organisms, we think, had more incisors than molars—this is common in Cretaceous Therians both in metatherians and eutherians. Yet the fossils Patterson found were largely constituted of molars and these far outnumbered the incisors that were found. Why might that be? It could be that when people are looking for fossils, molars are easier to see than are incisors. It might also be that molars persist better than do incisors but why would that be?

This leads us to recognize which of the teeth of a particular animal that died some 110 myo survive over a period of time is something that itself changes over time. At one time, close to the point of the death of the animal, the incisors would have predominated because these animals had more precanine teeth than molariform teeth, but as time wore on, a bigger and bigger percentage of the persisting teeth were represented by molariform teeth. We could provide an explanation of why these objects tend to persist differentially in their physical environment by making appeals to their various properties, which ground certain dispositions. We would mention the hardness of the teeth and that their shape, particularly that of the molars, is such that breakage is less likely. In addition, notice that the comparison class are not objects in general but rather a very special class of objects, the other alternatives available. In this case, the other teeth possessed by the animals.

In providing such an explanation of the character of the objects that persist in the fossil record, we would be providing a historical explanation of why teeth and particularly teeth of a certain shape, molariform rather than precanine, tend to persist in what is left of the early Cretaceous fossil record. We explain why the evidence presents itself to us this way in terms of the features of the objects that ground dispositional properties of the objects in particular to persist by resisting breakage, etc. Moreover, we are saying that these sorts of features of the objects account for their persistence in particular sorts of ways. When we use the theory of natural selection, this is the sort of explanation we always seek of the natural world.

In the case of the persistence of the molariform teeth, our explanation would be mistaken if it turned out that the real reason that these teeth persisted is because generations of ants gather objects of particular shapes including the shapes that these teeth happen to have and so save them from breakage.* In other words, our explanation would be mistaken about *why* the teeth have persisted even though we may be right that the teeth are in fact such that they are made of harder material and their shape is such that they are less likely to break through shearing pressure and so on. These facts may indeed ground a disposition to persist but unless their actual persistence depends in the right way on these features, our explanation of their persistence in terms of these features is mistaken. We may have thought that it was because of

---

* Intriguingly, ants do gather molariform teeth, as witnessed and discussed in the work by Bryan Patterson on the Trinity Formation from the Early Cretaceous near Forestburg, Montague County, TX (Patterson, 1956).

the features we nominated working in the way that we nominated that these teeth persisted. In fact, we can well imagine that we might have been wrong about why the teeth persisted. As we articulate the theory of natural selection, we must ensure that we maintain this way of getting things wrong. In general, when we are trying to give an account of a matter, making sure we can get things right and wrong in ways familiar to us is an important way of anchoring our account to the world.

The only way that we can make room for that sort of dependence of persistence on various features is if we allow ourselves to talk about either causation or, what may or may not be the same thing, counterfactual dependence. What we want is to be able to explain how it is that the ones that do persist do so because of their characteristics. That is a historical claim and it is a causal claim. The question of making sense of that "because" is what is at issue. Examples, such as the ones we have seen, support that contention. It might even be that although the shape of the teeth is the relevant factor, we were wrong about why the shape of the teeth was relevant. Therefore, we can imagine that the ants gathered the teeth into their nest thereby protecting them because the teeth had a shape similar to that of a seed which is used by the ants as a food source. In that case, the shape is indeed involved in the explanation of the persistence of the teeth but not in the way we nominated. We might have thought it was the way the shape resists breakage in terms of a shearing pressure that explains why the molariform teeth in fact persisted better than did the incisors. It could, however, have turned out that the shape was indeed the crucial factor but it turned out it was the way the shape resembled that of a certain kind of seed rather than the way the shape resisted breakage. In that case, when we sought to explain what led to the prevalence of molars in the present fossil record, we would be wrong if we thought it was because the molars resisted breaking better than did the incisors. We could be right that they resisted breaking and yet wrong that that is what explains their prevalence in the fossil record. This is a way we know we can get this sort of explanation wrong. When we present the theory of natural selection, or indeed the more general theory of persistence, we need to make sure that the theory doesn't just get the cases where we go right correct but also the ways we know we can go wrong. That is, as we articulate the theory, it has to be vulnerable to going wrong where we know we can go wrong. It is not a virtue of an articulation of a theory that it cannot go wrong where we know we can go wrong in the application of the theory. If for example, we would have been wrong to cite the shape of the teeth as explaining the molars' persistence because we thought it was their resistance to breaking that mattered when in fact it was the way they looked like seeds, then the way we articulate the theory must also make this sort of mistake.

The proto-theory of persistence actually plays a role in explaining which varieties tend to persist by nominating how that persistence depends on certain properties. The explanations we give are not at all guaranteed to be correct. We can go wrong by nominating the wrong sequence of persistents and we can go wrong by nominating the wrong properties as playing the explanatory role. Indeed, we can nominate the right property and get the way that property played the role in persisting wrong. When we use that theory we are doing history. We are trying to explain what the results we see before us. We are saying, given that these persisted, how is it that they did? What goes for this more general theory goes for explanation in the biological case too. This is exactly the way Darwin's key argument sought to explain how varieties tend to persist in a population by adverting to the advantageous variations that help them in their struggle to survive. Darwin's argument is interestingly simple and powerful. In fact, it is so simple that it is readily misunderstood. The connections with the more general theory of persistence are obvious too. The key argument does provide the central case for Darwin's mechanism and it looks like a very good case. What was remarkable about Darwin's argument is that he used information that was known to be facts by all and sundry (Huxley, 1901, 183). Facts, we could say, that were hiding in plain sight. He was able to put these well-known facts to good use in dealing with the problem of explaining both adaptation and diversity.

What differences are there between Darwin's biological version and the proto-theory of persistence? Darwin's theory uses the fact of heredity to give an account of how variations persist. The proto-theory of persistence does not actually spell out what persistence amounts to. It merely specifies that different varieties differ in their tendency to persist. The type of things that these varieties are is also left open by the theory of persistence. It is different for a glass to persist than it is for a soccer team. For a glass to persist, it itself must survive intact. For a soccer team, its survival can be constituted by being made up of different people over time. Moreover, in fact, Darwin presented his theory as applying to individuals. However, individuals need only be things that have descendants who carry heritable variations and which differ among themselves as to their fitness, or tendency to survive and produce descendants. Darwin, at one point, contemplated considering a beehive as a single individual. Later, we will consider what it is that bears the fitness and gets selected. For now, we can leave this question open.

Lewontin (1985, p. 76) suggests the following:

> Sufficient condition for evolution by natural selection is contained in three propositions:
>
> 1. There is variation in morphological, physiological and behavioural traits among members of a species (the principle of variation).
> 2. The variation is in part heritable, so that individuals resemble their relations more than they resemble unrelated individuals and, in particular, offspring resemble their parents (the principle of heredity).
> 3. Different variants leave different numbers of offspring either in immediate or remote generations (the principle of differential fitness).
>
> [A]ll three conditions are necessary as well as sufficient for evolution by natural selection.

Many people cite Lewontin and seem to suppose that he does give an account of Darwin's principles (Godfrey-Smith, 2009). It is important to notice both that Lewontin's three phenomena are different from the phenomena Darwin uses but also that Lewontin's proposition in no way entails natural selection. First, let us see how these propositions differ from Darwin's in a number of significant respects. The principle that "Different variants leave different numbers of offspring either in immediate or remote generations," although Lewontin calls this proposition "the principle of differential fitness," it is no such thing. It is really just a statement of the fact that different variants have different reproductive success. This is not equivalent to the principle that there is variation in degree of fitness because Lewontin's principles do not mention the difference in fitness. Calling differential reproductive success differential fitness is, as we have seen, a significant mistake. The mere fact of differential reproductive success is not a criterion for a difference in fitness, although we would certainly treat such difference as evidence for a difference in fitness. Just as we would treat the fact that one coin that came up heads more often than we would expect over a long run as evidence that the coin was biased. As Darwin insisted, and Lewontin ignores, fitness is the measure of a tendency to survive, not a measure of actual survival.

It is conceivable that we have all three of the phenomena Lewontin lists and still not have natural selection. Consider a population with many heritable variations, many differential reproductive successes, but in which that heritable variation is selectively neutral. In this scenario, Lewontin's phenomena are all present; however, natural selection will not explain the differential reproductive success of different variants because *ex hypothesi* the variations are selectively neutral. This situation, whose mere possibility is enough to show the limitations of Lewontin's formulation of the theory of natural selection, is one suggested by Kimura (1983) and Crow (1986) as closer to the norm than the standard neo-Darwinian picture of selection as omnipresent. We do not need the neutralist view to be correct to show that Lewontin has mischaracterized natural selection. The mere possibility that the neutralists are right is enough to show that the propositions he lists do what he says they do. Thus, he is wrong to assert that his three phenomena provide necessary and sufficient conditions for evolution by natural selection.

REFERENCES

Crow, J. F. 1986. *Basic Concepts in Population, Quantitative, and Evolutionary Genetics.* New York: W.H. Freeman.

Darwin, C. 1859. *On the Origin of Species.* 1st ed. London: John Murray; 2nd ed., 1860; 3rd ed., 1861; 4th ed., 1866; 5th ed., 1869; 6th ed., 1872; 6th ed., with additions and corrections, 1876.

Godfrey-Smith, P. 2009. *Darwinian Populations and Natural Selection.* Oxford: Oxford University Press.

Huxley, L., ed. 1901. *The Life and Letters of Thomas Henry Huxley.* Vol. 1. London: Macmillan.

Kimura, M. 1983. *The Neutral Theory of Molecular Evolution.* Cambridge: Cambridge University.

Kropotkin, P. 1902. *Mutual Aid: A Factor of Evolution.* 2009 paperback ed. London: Freedom Press.

Lewis, D. 1973. Causation. *Journal of Philosophy* 70(17): 556–567.

Lewontin, R. 1985. Adaptation. In *The Dialectical Biologist*, by R. Lewontin and R. Levins. Cambridge: Harvard University Press.

Patterson, B. 1956. *Early Cretaceous Mammals and the Evolution of Mammalian Molar Teeth.* Vol. 13, No. 1. Fieldiana: Geology, Chicago Natural History Museum.

Peirce, C. S. 1877. The fixation of belief. In *Philosophical Writings of Peirce*, edited by J. Buchler, 1955. New York: Dover Publishing.

5 Explanation, Causation, and Counterfactuals

NATURAL SELECTION AS EXPLANATION OF EVOLUTION

So far, we have seen that the circularity argument poses a serious challenge to Darwin's theory. Were the circularity argument to succeed, the theory Darwin proposed would be empty and would have no explanatory power at all. This becomes more pertinent when we notice that so many modern formulations of the theory do define key terms in the way that leads to the problem and so actually do strip the theory of any explanatory power. This means that we cannot define fitness in terms of actual survival numbers. Variations in survival rates can be due to numerous factors, and one of them is differences in fitness. Crucial to responding to the challenge so posed then is to distinguish between surviving and having different tendencies to survive. With that distinction in hand, it is clear that it is not trivial to state that the fittest survived. This is no more trivial than it is to say that the most fragile glasses broke; they tend to break, but it is not trivial that the ones that tend to break did break. They might not have broken after all. First, it is not trivial that any given glass is the most fragile: This depends on properties that it has and that we can investigate. In different contexts, different properties become important. What is most fragile in one context might not be in another; this brings in relativity to environment. We also saw the second and very important feature of explanations by natural selection in stark relief in the proto-theory of persistence: knowing which properties explain persistence is not enough, we need to know why they explain persistence. This point was illustrated by the persistence of molars in the Cretaceous therian fossil record. Knowing what we know about the mechanical properties of teeth, we can come to the view that molars resist breaking better than incisors. Their shape gives them a greater tendency to resist breakage over incisors. As we see the prevalence of molars in the fossil record, we could readily think that the feature we nominate, which we can suppose does indeed help them to resist breakage, explains their persistence. Patterson (1956) noted that molars may be collected by ants because their shape is similar to seeds, and they store them in their nests. This protects the teeth from surface exposure. Suppose he was right about this, then the shape did explain their prevalence but now not just because of the shape being suited to resist breaking. When we give explanations for persistence, we are offering causal explanations about the history that led to the persistence. We are specifying that on which that persistence (or more accurately, what the differences in rates of persistence) depended.

In this chapter, we turn to that relation of dependence. What exactly is involved in it? How do we seek to identify what depends on what? We will answer these questions as clearly as possible so we can be in a position to spell out what sorts of reduction are possible. The discussion so far has been abstracted from some very important details: What is it that is selected? What is it that fitness applies to? Spelling out Darwin's theory was done by making individuals in which we find differences in fitness and also what survives or not.

One aspect of the theory of persistence is that the form of explanation Darwin used is far more generally applicable. What that form of explanation requires is a difference in tendencies to persist (those differences are grounded in the properties of whatever we are considering), and we need a notion of persistence itself. The idea of persistence is not so straightforward as it might seem.

Richard Dawkins in his conceptual remarks on evolution makes persistence a key property of what he calls "replicators." However, Dawkins never actually spelled out what persistence is. He does say that genotypes change too much to be replicators, and so do chromosomes. He never considers phenotypic traits as such, not complete phenotypes but particular phenotypic traits. Dawkins allows that his favorite replicators do not persist by themselves persisting but rather by causing copies of themselves to be created. Moreover, that causal pathway can be quite indirect. Whenever that account works, there will be a trait that persists in just the same way. A trait, especially a heritable trait, persists by causing (albeit indirectly in many cases) a copy of itself. Then any heritable trait meets Dawkins's criterion for being a replicator. This was a bit quick; we will see later that the notion of heritability, which is crucial for Darwin's theory, has features that might lead us to worry about Dawkins's genetic reductionism.

We saw that the problems with the circularity argument could be avoided if we could define fitness in terms of probability of survival rather than survival. This option failed because we could not find a notion of probability that would play the role we need: it needed to be suitably independent of actual survival numbers but also explanatory of changes in survival numbers. In the end we had to reject the suggestion that we could define fitness in terms of probability. The proposal that met theoretical needs was to define fitness in terms of tendencies, dispositions, or potentials. Dispositions or tendencies do play an explanatory role in the behavior of their bearers. However, dispositions and tendencies are not without concern. They have a modal character that sits ill with a certain empirically minded approach to science. The disposition to break might be exhibited in breaking, but it is possessed even if the bearer does not break. This means accepting properties that are not directly observed but which play an explanatory role. For the empirically minded, talk of modalities is to be avoided if at all possible. This raises the possibility of characterizing natural selection and fitness in terms of conditional probabilities.* The problem is not limited to dispositions themselves but extends to causation itself. Causation is itself often surmised and not observed. However, we have seen that the theory cannot be so reformulated. There is an example that shows this rather well. Consider a population of organisms on its idyllic tropical island. Suppose that we have the knowledge of the distribution of fitness. We can even conceive of fitness as an objective chance of survival. Suppose we also know all the conditional probabilities that are not

---

* This suggestion was put to me by Bas van Fraassen when an early version of the material in this book (including the antiprobabilistic modal interpretation of fitness) was part of the work I did as a graduate student at Princeton University in 1990.

conditionalized on zero probabilities, if any such there be.* Now consider this happens: the fittest are observed to survive better than the less fit. Things go in accordance with the objective chances of survival we learned earlier. The question is, Can we learn from the observed rates of survival that the differences in survival depend on the features that the fittest have that make them the fittest? No, clearly we cannot. We can say that the fittest survived; the numbers justify that. However, we cannot say that the reason they survived is because they had the features that the fittest have.

The proto-theory of persistence was a generalization of Darwin's theory of natural selection, and the argument for it was a generalization of Darwin's key argument for natural selection. What the proto-theory of persistence displays in a particularly clear manner is the role characteristics play in persistence and how persistence depends on those characteristics. This idea of dependence needs teasing out.

NATURAL SELECTION, CAUSATION, AND COUNTERFACTUAL DEPENDENCE

Following Lewis (1973a), we can define a relation between events A and B. Let us say that B counterfactually depends on A exactly when the following two conditions hold:

1. If it were that A occurs, then it would be that B occurs.
2. If it were that A did not occur, then B would not occur.

This is an interesting relation on events. It has a modal character. When B counterfactually depends on A, this has consequences for what nonactual possibilities are like. Lewis (1973b) in his book *Counterfactuals* offers an account of when such sentences are true. This account is phrased in terms of overall similarity of worlds and states that would be true just when the most similar A world to the actual world is also a B world. From the time Lewis proposed this account, it met serious objections. Some related to the machinery of possible worlds. Some related to the particular way that Lewis's semantic clause worked. Kit Fine (1975) argued that Lewis's theory fails to give a suitable analysis of similarity and that on an intuitive understanding of similarity, Lewis's semantics give the wrong results. Consider the counterfactual, "If Nixon had pressed the button setting off an atomic apocalypse, the world would have been very different from the way it is now." Kit Fine suggests that this is intuitively true but the existence of such scenarios shows that Lewis's similarity based semantics for counterfactuals must be wrong. After all, there must be other scenarios where Nixon presses the button and something in such a scenario intervenes and blocks the apocalypse. Those scenarios are more similar to our world than are apocalyptic scenarios.

---

* Alan Hajek (2003) has argued that there are plenty of possibilities to which we have to assign a zero probability; possibilities that make perfect sense to conditionalize nontrivially. For example, the probability that any particular point is where a dart will land in a circle is zero. We could take conditional probabilities as primitive and assign values to probabilities conditional on zero probabilities in which case some of the standard equivalences would not hold.

David Lewis (1973a) thought that counterfactual sentences of the form "If it were that p, then it would be that q" allow us to make sense of causation, taking his lead here from an early but suggestive remark made by David Hume. Hume is most often associated with a regularity theory of causation: A type events cause B type events that follow them just when all A type events are followed by B type events. However, having characterized the regularity theory, Hume (1748, Section VII) says, "Or, in other words, where, if the first object had not been, the second had never existed."

Lewis defined a relation on wholly distinct particular actually occurring events, which is A causing B just when either B counterfactually depends on A or there is a chain of events (E_1, E_2, ..., E_n) between B and A, where E_1 counterfactually depends on A, E_2 counterfactually depends on E_1, ..., B counterfactually depends on E_n.

Famously, Lewis also advanced a semantic theory for counterfactuals that used the then surprising ontology of possible worlds and a relation of overall similarity on those possible worlds. Both of these features of Lewis's semantics invited significant debate. In fact, we can ignore that part of Lewis's views because the counterfactuals we accept are a better guide to the nature of similarity at work than our intuitive understanding of similarity is a guide to which counterfactuals are true. This suggestion also shows why the Kit Fine scenarios do not undermine the basic strategy of Lewis's approach: they trade on an intuitive grasp on similarity. It is a further issue whether there is a relation on scenarios that marches in step with the judgements we make about the truth of counterfactuals and that deserves the name of "similarity." Our substantive views about counterfactuals would seem to delineate such a concept at least to a workable degree.

The theory Lewis advanced in this paper changed the debate about causation, but it did not convince everyone. Some of the objections raised seem to have confused the notion Lewis sought to define with another notion the idea of A being the cause of B. In any case, Lewis himself moved away from this reductive story about causation.

The counterfactual analysis of causation Lewis provided does not succeed as an analysis of causation. We can see this most clearly when we look at the problem of preempted causes. It is an unavoidable consequence of his theory that there cannot be three actual events (with no intermediary events) e_1, e_2, e_3, in which e_1 causes e_3, e_2 does not cause e_3 but had e_1 no occured, then e_2 would have caused e_3. In such a case e_2 is a pre-empted cause of e_3; it is pre-empted by e_1. Whenever we have a preempted cause, then Lewis has to say that these three events are not immediately related. There must be at least one other event such that e_1 causes the existence of this event and that event blocks the causing of e_3 by e_2. Such an intermediary event would pull these cases into Lewis's form of explanation, but there does not appear to be any reason to think that whenever we have preempted causes, they necessarily do fit that pattern.

However, the problem the counterfactual analysis has with preempted causes, which undermines its standing as a theory of causation, is just the feature we can exploit in our explanations in the theory of natural selection.

The intuitive idea is this. Suppose we have an individual that is F and were it not F would have been G. Suppose further that we think that being F is what caused this individual's survival. That is, in actual fact, the individual is F, and that caused its survival. However, had it not been F, it would have been G (which it actually is not), and it would have survived as well. Thus, being F caused this individual's survival, but its survival is not counterfactually dependent on its being F. If it had not been F, it would still have survived for it would have been G, and that would have led to its survival. Rather, its survival is counterfactually dependent on its being either F or G. Thus, G operates like a modal alternative preempted cause: If it were not F, it would be G, and if it were not F, it would still survive because it would then be G, and that would cause its survival. Suppose this were a case of natural selection. What is being selected, being F or being F or G? Its survival is caused by its being F. However, its survival does not depend on its being F, but rather it depends on its being either F or G. Thus, the right answer would seem to be that what is being selected is its being F or G.* For it is because its being F is a way of being F or G that it survives. If it had not being F, *ex hypothesi*, it would still have survived because it would have been G. Thus, it is not its being F that explains its survival but rather its being F or G. Being F is in the causal chain that leads to its survival, but being F is not what its survival depends on. The role of natural selection is tied to the feature of the organism, which explains its survival, then that feature in the present case is its being F or G. Hence, although its being F is what causes its survival, what its survival counterfactually depends on is its being F or G.

NATURAL SELECTION AND FUNCTIONAL EXPLANATION

In philosophy of mind, we are still concerned to some degree with the problem that taxed Descartes: the problem of justifying the claims we want to make about our cognitive capacities. Like Descartes, there is the thought among some modern thinkers that we must look to our origin to find such justifications. Some (like Ruth Millikan) think that we can appeal to natural selection to provide the sort of account, which shows that our cognitive capacities have the sorts of properties were hoping for. Another much more ambiguous figure on the landscape of philosophy of mind is Dan Dennett. We will address Dan Dennett's philosophy of mind. Dennett is notable for his brave endorsing of the adaptationist stance that and Gould and Lewontin (1979) attack. In *The Intentional Stance*, Dennett (1987) says several things that constitute a fascinating, if hard to pin down, theory. Central in this theory seem to be the thoughts that there is something deeply right about interpretation having a constitutive role to play in a theory of the mind, and that such an account is not merely instrumental but objective. Moreover, there is the pervasive idea that intentional explanations and functional explanations provide "free-floating explanations." These views are not all that new. Donald Davidson might be justly thought of as having some paternal rights here. Dennett's own way of putting his claims is less than transparent although eminently readable. The problem seems to be less working out what he means in any given sentence than trying to find a way for him to end up being coherent. One of these problematic remarks of Dennett's is the suggestion that we should see evolutionary accounts of phenomena as a special case of the intentional stance, and that patterns in behavior are what the intentional stance used to explain

---

* This does not mean we are committed to any particular analysis of counterfactuals. Rather, it is to notice that this is how we use them in describing various cases.

in the end, patterns that are subject to explanation by the theory of natural selection (Dennett, 1991). All seems fine and clear up to this point. One is led naturally to the idea that far from being ambiguous on the instrumentalist/realist front, Dennett is a realist who, like Millikan, uses the theory of natural selection to underpin his account of intentionality. This is where things get complicated. Dennett gives us an account of proper uses of the theory of natural selection, which only makes sense read instrumentally. The worry is that this would infect his account of intentionality. Dennett is led to this conclusion by his views about natural selection and functional explanations. Dennett makes a lot of the fact that these were said to be "free-floating explanations." The only way they can be free floating is if the actual details of the causal history do not matter, and that means that the result would have been the same regardless of the competition. In that case, the organisms whose features are being explained are optimal from an adaptive point of view. Dennett seems to think that it is only if this is so that functional explanations can work.

There are good reasons to think that Dennett is wrong when he says that "adaptationist thinking in biology is precisely as unavoidable, as wise, as fruitful—and as risky—as mentalistic thinking in psychology" (Dennett, 1987, 283). In what follows, I will end up agreeing much more with notable critics of Dennett such as Stich (1981). Adaptationist thinking is the attempt to see the biological world as the outcome of natural selection operating alone and unimpeded through the millennia to yield organisms whose features are optimal. Although looking at biological features as having a function is crucial to making sense of the biological world, it is vital that giving functional explanations in biology is not committed to taking up an adaptationist stance. This means giving up the hope of "free-floating rationales." Moreover, the analogy Dennett draws between adaptationism in biology and mentalistic thinking in psychology is mistaken. The real analogy to draw is between adaptationism and a kind of ideal rationalism in psychology that attempts to explain everything people can be interpreted as doing in terms of beliefs and desires either explicit or implicit. Both are mistaken endeavors. Adaptationism as a theory of the biological world stands well refuted on various grounds, theoretical and empirical.

OPTIMAL CREATION

When it comes to trying to find justifications for our cognitive powers, some have thought that because evolution by natural selection is a satisficing process and not an optimizing one, wherein we cannot expect to use this theory to provide a justification for the claim that we get the world right. Something like this line of thought can be found in Descartes's *Meditations*: "But whether they ascribe my attaining my present condition to fate, or to chance, or to a continuous series of events, or to any other cause, delusion and error certainly seem to be imperfections and so *this ascription of less power to the source of my being will mean that I am more likely to be so imperfect that I always go wrong*" (Anscombe and Geach, 1954, 64).

The idea I want to focus on is that if I do not have my origin in an omniscient all-powerful and benevolent creator then it is likely that I am so constituted that I get everything wrong. Of course, in the context of the *First Meditation*, Descartes is arguing from the fact of his having fallen into error to the likelihood that his origin does not fit with a conception of God as all perfect. This possibility is something that he is at pains to dispel in the fourth *Meditation* by showing that his having fallen into error is his own fault.

It does not really matter whether the envisaged creator is a god or natural selection. We will see that no sensible account of intentionality or anything else had better depend on the idea that natural selection is optimizing in the sense to be specified because that idea is flatly false. The real question is to give an account of what follows if this creator, say natural selection, behaves in a less than optimizing manner. This is what I want to get to in the end. We have already gone some way to spelling a conceptual structure for explanation by natural selection that does not depend on optimality. The question remains whether optimality is required to give an account of functional explanation.

The simplest observation that shows that optimality is not viable to believe in is one of the key principles used in Darwin's key argument for natural selection and acknowledged by Huxley as widely recognized even before Darwin showed what followed from it: variation. In any population, there is significant variation. Therefore, not all varieties can be optimally adapted.

FUNCTIONAL EXPLANATION
AND COUNTERFACTUAL DEPENDENCE

We use the locution "for the sake of" in discussing biological function. It is interesting that the notion of function has not played any role in making sense of evolution by natural selection. However, we can quickly suggest an approach to biological function that sits well with the perspective we have been exploring. First, let me distinguish two distinct positions on function. The first is a selectionist view (Wright, 1973, 1976; Neander, 1991; Godfrey-Smith, 1993, 1994). A biological function of a phenomenon is the reason why that phenomenon was selected. This sort of approach has become the consensus view of biological function and has been extended into theories of meaning particularly by Millikan (1984).

Another approach was developed by Cummins (1975), where functions are effects that contribute to the explanation of capacities or dispositions of the systems they belong to. Cummins emphasized the arbitrariness of the capacities involved by emphasizing the role of the interpreter in deciding which feature to nominate and therefore which properties are functions will be subject to the selection of capacity explained.

Here I want to outline a third approach. This is the idea of the adaptive notion of function. A function of a feature is the role it plays in making the bearer of the property adaptive or, equivalently, in enhancing the disposition to survive. This can occur at any level; as long as we can specify what survival means, we can find attributes that enhance the capacity to survive.

First, let us see what is wrong with the selectionist account. It is certainly true that we could define a notion in this manner, but it will not be a very useful notion. According to that account, it is not that natural selection explains the origin of functions but rather that it is a part of the meaning of something being a function that has its origin in natural selection. In that case, it should not have been possible for anyone who did not believe in natural selection to identify what the function of a feature was. This is the same mistake as thinking that the notion of an adaptation, one of the two phenomena Darwin was trying to explain, should be defined in terms of natural selection. Darwin and everyone else understood the problem of explaining adaptation long before Darwin had thought of the theory of natural selection. It is not part of the meaning of an adaptation in that it has its origin in natural selection. It is a significant and contingent truth about our world that natural selection explains the origin of so many adaptations.

The adaptive notion of function identifies functions as those aspects that are adaptive. This can lead to selection, but whether the features are adaptive clearly does not depend on that. This is a restriction of Cummins's account. Not just any capacities are relevant to biological function. Contribution to fitness is the crucial marker of functions. This is agreed by the selectionist account and the adaptive account. Where the selectionist goes wrong is thinking that it is a part of the definition of a function that they are selected. Rather, we think it is a big part of the explanation of functions that they are selected, but it is not a part of their definition. The important thing is that with the theory of natural selection, we have an explanation of the origin of functions. It is contingent that functions have their origin in natural selection. They might have had their origin in a benevolent omniscient omnipotent creator or a bumbling but lucky creator. In any case, if organisms are as they are, then their functions would be the same. The contrast is sharpest with the selectionist account. For the selectionist, in neither of those cases would organisms have functions. For the adaptive account, what settles whether a feature is a function is its contributions to fitness. This is not easy to determine, but it readily explains why moving blood is a function of the heart and having the beat is not. It should also be noted that there is an implicit contrast: if not that feature, what?

When we come to explaining the origin of features that are adaptive, we are trying to specify what the origin depended on. One way to do that is to allow ourselves to talk about both causation and, what may or may not be the same thing, counterfactual dependence. The examples I gave will tend to support that contention. It might be that although the shape of the teeth is relevant, we were wrong about why it was relevant. In fact, I think counterfactual dependence is the useful notion. I do not think the counterfactual analysis of causation is successful in general, and that is because of preempted causes. I maintain the possibility of a world that consists of just three events related as cause, effect, and preempted cause. This is just not possible according to adherents of the counterfactual theory. However, it is just this feature that is exploitable in the theory of natural selection: Although being F caused this individual's survival, its survival is counterfactually dependent on its being either F or G. This seems to be the right answer to give if we are asked what is being selected for in this case. This does not mean we are committed to any particular analysis of counterfactuals. Rather, it is to notice that this is how we use them in describing various cases. When we say that functions are adaptations there because they are adaptations, the intuitive reading of the "because" is counterfactual dependence. If the organisms, considered individually, had not had the features that made them the fittest, they would not have survived.

Consider the Louisiana swamp hog we saw discussed by Darwin. These hogs are black all over and live in the dark shady swamps; moreover, they do tend to forage at night. It might be thought that this coloration is due to its effect as camouflage. As it turns out, the direct cause of the coloration is the pigment produced as a by-product of their detoxifying a plant commonly known as paint root. Domestic pigs no less than their feral cousins turn black when they successfully detoxify this food source. As it is, the black coloration is not much good for anything, but if any pig lacked it, it means the pig could not detoxify the paint root and so would be considerably worse off. Is the black coloration an adaptation? We defined that to be a feature such that the fitness of the organism is reduced if it lacks that feature. Well what if a pig lacked the black coloration, does that mean the pig could not detoxify the paint root and so would be worse off? If so, then the black coloration is an adaptation.

The "because" cannot be interpreted as counterfactual dependence if this example has been accurately described. In that case, the test counterfactuals are true for the black coloration as much as for the detoxifying ability. Does David Lewis argue that the color could have been lost without the detoxifying effect being lost in this sort of case? That would save the story but at the cost of some plausibility. It seems false that if the pigs were not black, they would still be able to detoxify the paint root. This shows that there is a problem with epiphenomena in David Lewis's story of causation in the case where the closest way of being without the epiphenomenon is to be without the cause of the epiphenomenon. In this case, the way most similar to the actual world in which pigs are not black has them being unable to detoxify paint root. Were David Lewis right, then there is no cause–epiphenomenon–effect triad such that if the epiphenomenon fails to occur, the cause and effect would both not occur. Moreover, this is something we can know a priori.

The issue of whether the transitive closure of counterfactual dependence is causation is not really pertinent here. What is at issue is whether we can ever find three events related in the way we have described. There is no logical reason for thinking that we will not find them. If we did find them, we would have to give up the counterfactual account of causation. However, that is no problem because we have already had to give up this theory because it rules out preemption of various simple sorts a priori.

Lewis can be right about counterfactual dependence even if he is wrong about causation and the semantics of counterfactuals. However, if the account was to be flawed, it would be because it counts things as selection, which are not. This gets remedied by increasing the sensitivity of the account to the modal relations involved. Because Lewis seems wrong about causation, we could use causation as a primitive notion in the story of functions distinct from counterfactual dependence. The causal role of black coloration is not relevant to its being there, although the causal role of its capacity to detoxify the paint root is. We then have a mixed theory invoking both causal elements and counterfactual dependence.

When we give an explanation involving counterfactuals in the way I have outlined, how optimal does the organism end up being? We have accepted that it is a part of the explanation that the variety that survived was superior to the alternatives it was competing with, and that is one sense of optimal: optimal relative to the selection class. This is not very optimal just yet.

We can make several useful distinctions when discussing optimality. We can distinguish, in what Sewall Wright called the fitness landscape, local maxima and global maxima of fitness. This point need not be labored; however, it is best not for-gotten. In addition, we can distinguish the optimality of the adaptation as the best solution to a particular "problem" set by the environment from the adaptation as the part of the best overall compromise of solutions to various problems set by the envi-ronment. The two distinctions we have made between overall optimal and optimal with regard to a given environmental problem as against the distinction between being at a global maximum and a local maximum are independent. Now the crucial thing to notice about "free-floating justifications" is that they depend on regarding the features before us as ideal solutions to the given problems we think they are solu-tions to. It is this insensitivity to the background of competitors and constraints that makes them "free floating." This sort of scenario never occurs. We know that we have to view the organism before us as a solution to all the problems it faces at once, with constraints put on it by the very fact that it grows and may well be faced with different problems at different stages of its development. Optimality might mean something weaker. Instead of being the global maximum of independent solutions to problems, we could say that the organism as we see it is the best compromise solution to all the problems at once. This is still a good notion of an organism being optimal, but there is no reason to believe that any organism is at the global maximum in its fitness landscape. Importantly, we still have not got free-floating rationales.

Dennett said that adaptive explanations were "free-floating explanations." The only way they can be free floating is if the actual details of the causal history do not matter. The problem is that sort of claim can only be sustained if organisms whose features are being explained are optimal from an adaptive point of view. Dennett does think that it is only if this is so that functional explanations can work. I hope to have convinced you that Dennett is wrong. We have seen that functional explanations are available for those who sensibly think the world is not that of the adaptation-ist. This means that we can rethink by analogy the rationality assumption implicit in Dennett's account of the intentional stance. Perhaps we should only ascribe just enough rationality for the intentional explanation to work. However, here we really do lack a theory of the sort Darwin provided biologists. The worry is that "free-floating rationales" in this context mean just that.

The upshot of this section is that the key role of natural selection is giving explana-tions about the way differential survival depends on characteristics. In deploying the mechanism of natural selection, we are making historical causal claims. These do not float free of the details of actual history. We are claiming that the changes we have observed are due to the fittest differentially surviving because they have those charac-teristics that make the fitter. This "because" expresses a dependence. There is a causal aspect—their differential survival is caused by their having those characteristics—but there is also something quite distinct, a counterfactual dependence, the "but for" story—but for those characteristics, they would not have differentially survived. The second is quite different from the first in that it implicitly brings in a comparison with some alternatives. Those characteristics cause the differential survival and in certain ways. To make such a claim is to make a causal historical claim. We can be wrong in all the ways we know causal historical claims can be wrong. We can get the timing wrong. We can know what happened when and still get the reasons wrong.

Evolutionary history is patchy. There is a lot of happenstance in our actual evolutionary history. There have been significant massive extinctions at the end of the Cretaceous, Triassic, and Permian, for example. Getting through such extinctions seems to be a matter of luck more than great adaptedness. At a smaller scale, many populations go through population bottlenecks. These have a significant founder effect on the future of that lineage should the population recover. When populations are smaller, the relative importance of stochastic effects such as random genetic drift becomes greater. This seems to have been important even in the evolution of our species with some suggestion that the population of sub-Saharan humans might have dropped to less than 2000 before starting to recover some 100,000 years ago (Hawks et al., 2000). The relentless process of natural selection nevertheless has an impact on the biota. No selective history compares any successful variety with all possible alternatives. There has been a particular course of history with a particular range of alternatives and, considering all that is possible, a relatively narrow range.

DOES CHANCE PLAY A ROLE IN DARWIN'S THEORY?

Darwin's theory of natural selection aims to explain the way organisms come to be adapted to their environment and, because those environments differ, how organisms that come from a common ancestor can diverge in character. The way the theory does that explaining has been the focus of this investigation. It has been said that Darwin's theory is the truly modern scientific theory. Stephen Jay Gould famously argued its novelty lies in the way it represents a wholly materialist attempt to explain the organic world. Gould probably overstates things here. Darwin actually does not commit himself to materialism in *The Origin*, and although the theory may be consistent with materialism, the theory does not require it. How else does the theory look so thoroughly modern? Another suggestion championed by Charles Sanders Pierce is that the theory is significant because it is the first theory to make central use of chance, and this is important because probability is the language of modern science. In fact, the role of chance in Darwin's theory can be quite misleading.

Does chance play a role in Darwin's theory of natural selection? If it does, what role is that? The first question seems obviously answered by many passages in Darwin—and commentators on Darwin—that emphasize the role of chance in his theory. Here is a typical quote from *The Origin*:

> Owing to this struggle for life, any variation, however slight and from whatever cause proceeding, if it be in any degree profitable to an individual of any species, in its infinitely complex relations to other organic beings and to external nature, will tend to the preservation of that individual, and will generally be inherited by its offspring. The offspring, also, will thus have a better *chance* of surviving, for, of the many individuals of any species which are periodically born, but a small number can survive. I have called this principle, by which each slight variation, if useful, is preserved, by the term of natural selection, in order to mark its relation to man's power of selection. (Darwin, 1859, 61; emphasis added)

Again when discussing the "mutual checks to increase," Darwin draws attention to the way what can look like chance may have law-like underpinning.

> When we look at the plants and bushes clothing an entangled bank, we are tempted to attribute their proportional numbers and kinds to what we call chance. But how false a view is this! ... Throw up a handful of feathers, and all must fall to the ground according to definite laws; but how simple is this problem compared to the action and reaction of the innumerable plants and animals which have determined, in the course of centuries, the proportional numbers and kinds of trees now growing on the old Indian ruins! (Darwin, 1859, 74–75)

In this case, Darwin is using "chance" to mean the same as "undirected," "random," or "unlaw-like." His argument is that we can readily explain what looks like chance to be actually perfectly law-like and not undetermined. The question is whether this applies to all applications of the notion of chance or not. After all, we saw that he seemed to use the notion of chance when talking of "better chance" of surviving in his characterization of natural selection.

Another very common way Darwin uses "chance" is when he is discussing the origin of variations. This is another typical expression:

> In such case, every slight modification, which in the course of ages chanced to arise, and which in any way favoured the individuals of any of the species, by better adapting them to their altered conditions, would tend to be preserved; and natural selection would thus have free scope for the work of improvement. (Darwin, 1859, 82)

This talk of modifications that chanced to arise, however, points to a different sort of use of the term "chance." The first was tied to what seemed unlaw-like, and Darwin was concerned to show that it may nevertheless be law-like. The second use, I argued, must be understood not probabilistically but rather dispositionally. The fittest have a stronger disposition to survive. Now this third use is quite interestingly different from the first two. The idea is that modifications arise by chance. It is clear he does not mean that there is a hidden law-likeness to the modifications. Rather, the idea is that they are independent of the selection pressure.

Darwin himself mentions that the advent of variation is due to chance, but more, in many cases, even the survival of individuals is described as fortuitous. So is the case closed? No, I want to convince you that this is an illusion. Chance, properly so-called insofar as it requires indeterminacy, does not play that sort of role in Darwin's theory. In fact, Darwin's theory does not depend on chance at all, and it is important that it does not.

Darwin does need the variation that arises in a population, and he emphasizes something that was not obvious to anyone before he pointed it out. However, before Darwin, the variation observed in a population was held to be noise—chancy, inexplicable, and irrelevant—perturbations about the ideal type of the population. The various lions we observe are contingent variations from the ideal form of a lion, for example. In Aristotle's explanations, every individual would be exactly realizing the type were it not for these perturbations in development. Actually, it is a bit more complicated than that even for Aristotle. For example, females were in one sense monsters because they deviated greatly from the form supplied by their fathers, and yet Aristotle knew that without females, there could be no sexual reproduction. It is just not clear how the ancient biologist and philosopher resolved this tension: Could there necessarily be contingent monsters?

What Darwin saw clearly was that the origin of heritable variation was independent of the selection process. This was all he wanted to say about the process. He saw that the scatter of characteristics about the norm seem random with respect to the particular selective pressures that presumably are operating. Even here, this deep and important insight is only true, as Aristotle used to say, for the most part. There are exceptions to Darwin's insights, and these exceptions serve to show just how much of the time his observations correctly describe the way things go with variations being unrelated to the selective pressure. So what sort of exceptions are these? First, there is the fact that sexual reproduction is optional in some species, and choosing that option as opposed to parthenogenesis (virgin birth) seems to be a way of increasing variation in the offspring. Conversely, choosing parthenogenesis seems to be a way of restricting variation, leading to fewer deviations from a successful model. Notice that in this example, sexual reproduction does not skew the direction of the variation but rather has an impact on the degree of the variation.

In fact, Darwin does comment in *The Origin* about mechanisms that may increase variation. In the key chapter on natural selection, he says, "We have reason to believe, as stated in the first chapter, that a change in the conditions of life, by specially acting on the reproductive system, causes or increases variability; and in the foregoing case the conditions of life are supposed to have undergone a change, and this would manifestly be favourable to natural selection, by giving a better chance of profitable variations occurring; and unless profitable variations do occur, natural selection can do nothing" (Darwin, 1859, 82). The central point though is that variation is independent of natural selection, "Unless profitable variations do occur, natural selection can do nothing."

Here is the claim: Despite all the evidence to the contrary, Darwin never needed to appeal to chance. What he needed was that variations do not come about for the sake of the advantage they confer. This is importantly different. That is why a world without chance, a deterministic world, can nevertheless be a world in which there is Darwinian evolution by natural selection.

There is a crucial ambiguity in the way we talk about chance. There is a well-known concept that has already figured as one of the possible interpretations of probability when we were looking to understand the notion of fitness. This notion is sometimes called "objective chance" or "propensity," and it plays a central role in quantum mechanics. Earlier, I argued that the notion of objective chance cannot play the role of making sense of the phrase "most likely to survive" in a manner that would allow it to avoid the circularity problem. Be that as it may, is there perhaps another role that the notion of objective chance has to play in the theory of natural selection? Notice the way I put the question. It is not whether the theory of natural selection is consistent with a world that has objective chances in it, but rather whether the theory of natural selection requires such a world.

Hence, if the quantum physicist's notion of propensity or objective chance is not required, what is? Suppose you have met a friend "by chance" as we say in a faraway town. New York City seems to be such a city for me. Whenever I go there, travelling halfway around the world, I seem to meet people I know and that I had no idea would be there. Certainly, we would respond to the question why I met those people perfectly accurately by saying it was happenstance or chance. When we explain things that way, we do not mean that it was a quantum indeterministic event. At base, it may well be such an event, but that is not what we mean. In fact, their being there is normally fully explicable by their intentions, in the same way as my being there will have been planned months ahead. What do we mean by the phrase "by chance" in that context then? Because each event is caused or even determined, the whole is caused or determined. What we mean, it seems, is that the meeting itself was not intended. It was a meeting that was determined but not as a meeting. My being at that place is caused (in large part by my intentions); their being there was caused (again in large part by their intentions). The meeting then was caused by these causal chains, but in neither of our intentions was there an intention that involved our meeting in that place. Thus, the meeting was caused by our intentions but not intended.

This notion of what occurs not being aimed at is what Darwin needs for his theory of natural selection. He needs that there is variation in the population and that new variations occur. These can be caused, and perhaps we will discover the various causes of variations in populations. We certainly are aware of significant explanations for a lot of variations. Suppose we discover all the various mechanisms that lead to variation. What Darwin was keen to point out was that such variation is not brought about for the sake of natural selection. In this detail, according to some theorists, he may have made mistakes. It may be that some features of organisms are selected because of their tendency to generate variation. One example of a characteristic that generates variation is sexual reproduction itself. Much of the variation we see in sexually reproducing species is not due to the advent of new mutations but rather due to the rearrangement of genetic material on chromosomes in a process called "crossing over" during the formation of gametes (meiosis). Crossing over in chromosomes seems to depend on sexual reproduction. Could sexual reproduction have a selective advantage in the very variation it introduces? That is one intriguing suggestion. Conversely, and on the other hand, there are various organisms that, when they hit on an adaptive form in an especially intensely competitive situation, revert to asexual reproduction, which has the effect of maintaining the adaptive phenotype.

Thus, objective chance needs to play no role for Darwin's theory of natural selection. His theory, as I have argued in Chapter 4, was developed in a context in which the underlying physics was understood to be deterministic. His theory in fact is neutral on the nature of that underlying physics over a large range of theories, not utterly neutral because he needs causation. Darwin needs that the sorts of explanations we give of natural selection that use the idea that some characteristic allowed a variety to survive and reproduce are possible, and for that to be possible, that talk of "allowing" must be sustainable.

What Darwin really needs is heritable variation. When he says that the variation is random or due to chance, he does not mean that it is not caused or determined by antecedent events, and he does not mean that it cannot ever be explained even in principle. What he means is that the variation is independent of and comes before the selection. The variation comes about howsoever that it does, and then selection occurs. The variation does not come about because of the selection. As we have seen, there are special cases in which this might be wrong, cases in which features in organisms arise that can modulate the frequency and amount of variation. Even in these cases, the variation is not directed: more variation can occur with sexual reproduction, but just which variations occur is not itself intended. Such cases just go to show how Darwin's observation is profoundly correct and how his theory generates the sorts of explanation he sought. Thus, Darwin's theory does not depend on chance, not the sort of chance we find in quantum mechanics, and to think that his theory is mortgaged to such chance is a fundamental mistake about the way his theory explains. Organisms can be lucky in their characteristics. We can say that they had the right features "by chance," but when we say that, we are like Aristotle's lucky farmer who finds a hoard of treasure by accident when digging a hole to plant a tree (Aristotle *Methaphysics* Δ, Chapter 30). The hoard was hidden intentionally. The hole was dug by the farmer intentionally. There was no intention to find a hoard of treasure; that was not the intention with which the hole was dug.

In making use of a notion of "chance" that was consistent with determinism and did not require genuine chance in the world, Darwin was following a tradition that reached back to Aristotle. Here is Boethius in a text written while he awaited execution in jail. It is written as though he were visited in jail by the personification of Philosophy, who appears as a woman in this dialogue:

> "Well," said I, "is there, then, nothing which can properly be called chance or accident, or is there something to which these names are appropriate, though its nature is dark to the vulgar?"
>
> "Our good Aristotle," says she, "has defined it concisely in his 'Physics,' and closely in accordance with the truth."
>
> "How, pray?" said I.
>
> "Thus," says she, "whenever something is done for the sake of a particular end, and for certain reasons some other result than that designed ensues, this is called chance; for instance, if a man is digging the earth for tillage, and finds a mass of buried gold. Now, such a find is regarded as accidental; yet it is not 'ex nihilo,' for it has its proper causes, the unforeseen and unexpected concurrence of which has brought the chance about. For had not the cultivator been digging, had not the man who hid the money buried it in that precise spot, the gold would not have been found. These, then, are the reasons why the find is a chance one, in that it results from causes which met together and concurred, not from any intention on the part of the discoverer. Since neither he who buried the gold nor he who worked in the field intended that the money should be found, but, as I said, it happened by coincidence that one dug where the other buried the treasure. We may, then, define chance as being an unexpected result flowing from a concurrence of causes where the several factors had some definite end. But the meeting and concurrence of these causes arises from that inevitable chain of order which, flowing from the fountain-head of Providence, disposes all things in their due time and place." (Boethius, *The Consolation of Philosophy*, Book V, Chapter I)

EXTENSIONALIST REDUCTIONISM, CAUSATION, AND EXPLANATION: THE CASE OF THE IDENTITY THEORY OF MIND

The most significant advocate for the extensionalist reductionist picture is David Lewis, probably the greatest metaphysician of the twentieth century. Certainly, from where we stand, less than two decades after his untimely death, he is still the most influential metaphysician for contemporaries. Lewis argued for a wholesale reductionism and for an extensionalism. Lewis provided an argument for the identity theory of the mind. Elsewhere, I have argued that the positive argument Lewis provides for the identity theory fails (Michael, 2013). The fundamental problem is a gap between arguing that physics explains everything and arguing that every explanation is at root a physical explanation. It was suggested that it is possible that there be explanations that are not reducible to physical explanations even while accepting that everything is physically constituted.

ARGUING AGAINST THE IDENTITY THESIS

Suppose we grant that the world is entirely constituted of physical stuff. In what follows, I shall argue that among the states of the world that explain various phenomena are mental states, and at least some of these are not physical states.

Take a very typical example of how explanation works in giving an account of our actions. Suppose that I lean forward because I believe that if I lean forward I will hear better what Lloyd is saying. Such an explanation is of course partial but serves well to show how beliefs figure in the causal explanation of behavior. In providing an explanation of a particular event, I indicate a chain of events linked causally to that event. In this, we are following Lewis (1986). In this particular case, I do this by nominating a salient member of that chain, my having a certain belief. Suppose that my actual leaning forward, f, is caused in this way by my actual belief, b, and that this belief is constituted, in fact, by my being in physical state p.* When we say that b causes (or causally explains) f, we imply *modulo* some subtleties, that if b had not occurred, then f would not have either; in other words, that f counterfactually depends on b.† Most of the subtleties relate to cases of causation without counterfactual dependence. These do not arise here because f does counterfactually depend on b. If I had not been in physical state p, another physical state would have occurred that would also have constituted my believing that I will hear better what Lloyd is saying if I lean forward. This is to say if I had been in a very slightly different physical state, I would nevertheless have been in the same belief state. Think of all the very many ways you can be slightly physically different without that affecting your mental state. If I had not believed what I believed, I would have been very different physically. However, this shows that physical states

---

* There is no assumption that the physical state that constitutes my believing as I do merely relates to my body. It may but it is not part of what I argue for or assume here. The physical state I am in may be rather extensive; it may be significantly more inclusive than what happens in my body.

† See Lewis (1973a). Note that it is no use for the identity theorist to say that although f does counterfactually depend on b and not on p, that nevertheless there is a chain of events, q, r, ..., z such that q counterfactually depends on p, r counterfactually depends on q, ..., and f counterfactually depends on z; thus, p and f are related by the ancestral of counterfactual dependence. However, this does not change the fact that f does counterfactually depend on b and not p. This shows the irrelevance to the present matter of the analysis of causation. If someone doubts the counterfactual account of causation, then the argument still goes through as an asymmetry of counterfactual dependencies of mental and physical states.

and the belief states they constitute embed differently in counterfactual and causal contexts. The mental state (belief *b*) and the physical state (*p*) it is constituted differ in their counterfactual properties; therefore, the physical state does not play the same causal role as the belief, although it constitutes it.

Thus, the argument against the identity thesis runs as follows:

Premise 1: The definitive characteristic of any (sort of) experience as such is its causal role, its syndrome of most typical causes and effects.

Premise 2: However, experiences and the physical states that constitute them differ in their causal roles.

Conclusion: Because those physical states do not possess the definitive characteristics of experience, they cannot be those experiences.

The belief (*b*) and the state it is constituted by (*p*) differ modally. We have seen that if *p* had not occurred, *b* would still have but have been constituted by a slightly different physical state. It is a question begging to respond to this argument with the suggestion that if *p* had not occurred, then *b* would not have either. It is simply incredible to think that the mental state we are in is as subject to the contingencies of the physical state that constitutes it. If some of my teeth had four less atoms in its crystalline structure, I would have believed just as I do. Physical states are individuated very finely. A difference in number of atoms is enough to make it a different physical state. However, this may not make a mental difference at all. Intuitively, total physical states and total mental states are just not similarly carved up. If *p* had not occurred, that is if I had not been in that particular physical state I was actually in, then I would have been in a (distinct but) similar physical state, which would also have constituted my believing what I actually do.

One strategy then is to assert that if I had been in a very similar but distinct physical state *p**, then I would have had a very similar but distinct belief *b** with the same content. As evidence, couldn't it be said that, after all, the state that constitutes *b** is distinct from the state that constitutes *b*? However, notice that we earlier assented to the fact that if I had not had the belief that I did have, I would not have leaned forward. On the suggestion that *b* is not *b**, we should have said that I would still have leaned forward, for then *b** would have underpinned that leaning. However, crucially, then we would lose the sense of my belief being a cause of my leaning forward.

Of course, another option is that there was the actual leaning *l* caused by *b*, but had *b* not occurred but been replaced by *b**, *b** would have led to another leaning *l**. This option fails as it seems what we are explaining in adverting to my belief that I will hear Lloyd better if I lean forward is my leaning forward at all. The explanatory contrast is between my leaning forward and my not leaning forward, not *l* occurring and *l** or its like occurring.

It is only by running together these two series of modal properties that the contingent identity thesis becomes at all plausible. Indeed, when we note that the causal role of a state is what makes it a mental state, and we note that a causal context is either a modal context or at least has modal implications, then any account of the nature of mental states that focuses only on their actual constitution will be radically mistaken. Our conception of mental states, when they are identified by their causal roles, is *ipso facto* modal. To earn the name of a mental state, a physical state must embed the right way in a series of counterfactuals.

Because our theory of the world posits causal relations and is thus modally involved, we should say that constitution is not in general identity.* It is an interesting and metaphysically substantial thesis to say that something that constitutes a thing is identical to that thing. It is not always the case. The argument provided shows that our usual commitments together with the principle of physical constitution entail that beliefs are not identical to the physical states that constitute them. This completes the argument for the thesis that token mental states are not all identical to the token physical states that constitute them. We have seen that these states do not embed in the same way in counterfactual constructions and so have distinct causal roles.

SKEPTICAL REACTIONS

CHANGE OUR UNDERSTANDING OF THE MENTAL?

The argument presented leads to the conclusion that a particular belief can be distinct from the particular physical state by which it is constituted. Someone might react by saying that this shows we need to revise our conception of mental states stripping off from them the modal commitments, which served to distinguish them from the physical states by which they are actually constituted. Such a move might be motivated by ontological economy. We have the physical states that constitute everything, why think that there are any others over and above these? This suggestion sounds ontologically economical, but we pay a theoretical cost. The key role of mental states is their characteristic role in the causal explanation of phenomena, including physical phenomena. We have already seen how such explanations can fail to reduce to physical explanations. There are many particular forms such a revisionary view might take, but none salvages anything recognizable as our notion of the mental, in particular, as it figures in the explanation of action. For that reason, this sort of revisionism is unappealing.

The attitude I am recommending, in contradistinction to this revisionism, leaves this part of our ordinary notion of the mental, and in particular its explanatory role, intact.† Thus, although the thesis of the physical constitution of mental states is an interesting, substantive, and contentious thesis, we give up the prospect of a physicalistic reduction by adopting it in preference to the identity thesis, which would serve to explain behavior. It was this prospect that made the identity thesis so interesting in the first place. As it is, we must see that apart from physical states, the world also has mental states. Although the physical states may constitute all the others, they are not identical with all the others.

---

* This traditional view has been endorsed in Kit Fine (1992) and in Mark Johnston (1992).
† It is as well to note that I am not here addressing that part of our notion of the mental that fits with the notion of a subjectivity, something which it is like to be. I am focusing exactly on the public role of mental states in explaining actions. Nothing I say here commits me either way on the notion of subjectivity.

But Causal Contexts Are Opaque

In defense of the identity thesis, it might be said that causal contexts are opaque so that coreferential terms may not be substitutable *salve veritate*; therefore, the argument I gave is invalid.* My argument suggested a difference in the causal properties of mental and physical states on the basis of the difference in the way names of these states embed in counterfactual contexts.† If the context is opaque, then names for the same thing may not be substitutable *salve veritate*. However, Lewis now needs a positive reason for asserting that mental and physical states have the same causal roles although names of these states embed differently in these contexts. Their similarity in causal roles was supposed to tell us that they are identical. Now we need to suppose that they are identical to smooth out the wrinkle that they appear to have different causal roles by dint of embedding differently in counterfactual contexts. As we have seen, to smooth out that particular wrinkle takes more than merely being someone who thinks that everything is constituted physically. Indeed, it takes having already identified mental and physical states. In the context of the present argument, that identification would be question begging because that is the conclusion of the argument, which is here in dispute. As for my argument, the fact that names of mental states embed differently from names of physical states in counterfactual and, therefore, in causal contexts can be taken at face value as evidence that they have different causal roles. In the absence of an independent reason to think that this phenomenon should not be taken at face value, it is very good evidence that they have different causal roles.

REDUCING "BEING UNLOCKED": A PARALLEL CASE?

So far, I have tried to show that Lewis's argument for the identity theory contains a crucial lacuna: even if there are physical explanations for all physical phenomena, it does not follow that all explanations of physical phenomena are physical. Further, I suggested that paying attention to the modal properties of beliefs and states that in fact constitute them shows that they cannot be identified. In the same way, the clay and the statue it constitutes have different modal properties and so cannot be identified. This might suggest that there is a wholesale denial of the possibility of reduction; that the modal argument I have deployed can be wheeled out to undermine any significant reduction. This is quite mistaken. To show that this is not so, we do well to consider the illustrative, uncontentious argument Lewis alleges is parallel to his argument. We will also discover an argument that might seem to undermine my own argument against the identification of mental states and the physical states that constitute them; the suggestion might be that the physical states can have a disjunctive character. This argument will be considered and rejected. However, it will yield some important lessons.

---

* Lewis did not discuss this issue to my knowledge. However, in Lewis (1971), he said that a multiplicity of counterpart relations makes for the nonsubstitutivity of coreferential names in modal contexts.
† Part of Lewis's discussion of this point would seem to depend on the view that some names are contingently referential. I think that there are no such names. Denotation can be contingent but not reference. Reference is our mode of tracking things through changes, temporal and modal.

Lewis (1966) provided what he considered an uncontentious example of onto-logical reduction to highlight the features he saw shared between it and the case of mental states. He says,

> Consider cylindrical combination locks for bicycle chains. The definitive characteristic of their state of being unlocked is the causal role of that state, the syndrome of its most typical causes and effects: namely, that setting the combination typically causes the lock to be unlocked and that being unlocked typically causes the lock to open when gently pulled. That is all we need to know in order to ascribe to the lock the state of being or of not being unlocked. But we may learn that, as a matter of fact, the lock contains a row of slotted discs; setting the combination typically causes the slots to be aligned; and alignment of the slots typically causes the lock to open when gently pulled. So alignment of slots occupies precisely the causal role that we ascribed to being unlocked (for these locks). Therefore alignment of slots is identical with being unlocked (for these locks). They are one and the same.

This example is instructive. Let us grant what seems plausible, that Lewis is right that being unlocked for these locks and having the slots aligned are one and the same. It should be noted that this does not lend credence to his contentious argument identifying mental and physical states because the states of being unlocked and hav-ing slots aligned do support the same counterfactuals. If the slots were not aligned, the gentle pull would not have opened the chain. If the chain were not unlocked, the gentle pull would not have opened the chain. Thus, far from being a counterblow, this argument fits exactly the *desiderata* for a candidate for reduction. This only makes more obvious the fact that Lewis's argument for identifying mental and physi-cal states is not analogous to this uncontentious case because at least some mental and the physical states that constitute them do not support the same counterfactuals. Assessing counterfactuals takes us to other possibilities. Going to possibilities in which the physical states are different might be different from going to possibilities in which mental states are different, even if actually mental states are physically constituted.

Although we have agreed with Lewis that the physical state of having slots aligned and the state of being unlocked support the same counterfactuals and so can be identified, there are states that being unlocked is constituted by and to which the state of being unlocked is not identical. The same sort of argument deployed against the identity thesis will apply here. Call the complete physical state that constitutes the lock's being unlocked now, *p#*. If the lock were not unlocked, it would be locked. However, if the lock were not in *p#*, it would still be unlocked for it would be in a physical state similar but not identical to *p#*. In other words, or in another's words, the closest physical states to *p#* would nevertheless be a state that constitutes the lock's being unlocked. In that case, if the locked were not in state *p#*, it would never-theless be locked. However, because the physical state of the lock and the state of its being unlocked have different counterparts, they are distinct states.*

---

* Thus, it also follows that what counterfactually depends on the lock's being in *p#* is not what counterfactually depends on the lock's slots being aligned. Thus, these two states are distinct.

Does this argument show that the state of being unlocked is a nonphysical state of the lock? No. Being unlocked is not a state of the lock, which is identical to the lock's being in $p\#$. On the other hand, the state of the lock's being unlocked is constituted by $p\#$ and by nothing else in the actual world. However, its being unlocked could have been constituted by any of several distinct physical states, not all of which are counterparts of one another.*

A response that might seem attractive at this point is to say that although the lock's being unlocked is not its being in $p\#$, it is the lock's being in some disjunction of states of which $p\#$ is one disjunct. I shall present two arguments against this position. The first is an argument from premises Lewis himself has granted, which points to the lack of a guarantee that this disjunctive state exists, and that even if it does exist, it is not obviously going to be available to explain what needs explanation. The second argument has to do with the character of the disjunction itself even if it is held to exist. This second argument is not a general argument. It is quite likely that this second argument is applicable in the case of mental properties and not in the case of the lock. In which case we have a reason to think once again that the cases Lewis holds as analogous are not, and instructively not, analogous. If that arises, we would have reason to think that the lock's being unlocked is the same as the lock's being in one of a range of physical states, whereas the same does not hold for the mental state.

Before considering these arguments, it is well to consider the general relation of disjunctive states and the states that constitute them. Suppose that $p\#$ involves having a property q and suppose that the lock does not have a distinct property r. Then being in $p\#$ constitutes that lock's having the property q-or-r. However, although its having q constitutes its having q-or-r, these are not one and the same. For if the lock were not q, it would be r. However, if it were not (q-or-r), it could not be r. In that case, clearly its having one property constitutes its having another but distinct property. The notion of distinctness among these properties is clearly not a logical notion but rather metaphysical. The relation among the properties may well be necessary and a posteriori and so not logical, but in some cases there may indeed be a logical relation among the properties. Another illustrative example: Suppose the property of running is the same as the property of running quickly or slowly. It may be that someone's running slowly is what constitutes their running. Now it is clear that what may depend on their running slowly may not depend on their running. For it might be that if they had not run slowly, they would have run quickly, whereas if they had not run, they would have walked. In that case, although their running slowly is what constitutes their running, it is not the case that what depends on their running also depends on their running slowly.

The issue of whether there is a physical state of being in some one of a group of distinct physical states is not in general an easy one to answer and depends in part on the cardinality of the set of states and their physical characterizations. Whether

---

* Note that something weaker might be obtained, which would do some of the time. There may be no physical predicate that gathers all the p_i, which could have constituted the locks being unlocked. However, there may be a physical predicate that gathers together all the physical predicates that would have constituted the locks being unlocked under some supposition. That is, there may be a predicate that gathers together all the physical ways the lock would have been under a given supposition. However, this too is far from guaranteed.

there is a predicate in the language of physics that appropriately maps our ordinary predicate "is unlocked" or "has slots aligned" becomes the issue of whether there is a predicate that gathers together the various physical states pi, which could have constituted the lock's being unlocked.* This would be settled in the affirmative if there is also the disjunctive state of those particular states whenever there are several physical states, or perhaps the existential state of being in some one of the particular states. However, this is not guaranteed by the mere fact that there is a predicate for each of the states to be gathered together. With regard to the analogous question to do with properties, Lewis (1986, 224) said in one case, "The existential property, unlike its various bases, is too disjunctive and too extrinsic to occupy any causal role. There is no event that is essentially a having of the existential property; *a fortiori*, no such event ever causes anything."

Thus, the existential property, or the disjunctive property, is not in general going to be available for Lewis to ground the causal dependence of the opening of the chain on the alignment of the slots. It is important to remember that the opening of the chain does not causally depend on its being in $p\#$, although the opening of the chain does causally depend on the lock's being unlocked. The opening of the chain does stand at the end of a train of events that are pairwise, one counterfactually dependent on the other. However, counterfactual dependence is not in general a transitive relation. This was noticed by Lewis (1973a) when he first introduced the notion. Of course, even if states are not in general closed under disjunction, it might happen that whenever we have a state that figures in a causal explanation, a corresponding disjunction of distinct physical states is also a state. This would ensure that the disjunction of the pi's is also a state available to figure in the reduction of the lock's being unlocked. However, the fortuitous happenstance that such a disjunctive physical state was always available in such situations, although not being available in general would be a truly remarkable coincidence, and it is not guaranteed as a logical consequence of the premise of physical constitution.

Turning back to the case of the reduction of the belief that causes my leaning forward to the underlying physical states which constitute it, we might on further reflection worry whether a reductive physicalist's disjunction should be regarded as an adequate explication of an actual state because many of the disjuncts are merely possible states. Does this make the belief a physical state but one which is partially identical with otherworldly possibilia? This does seem a bit odd, but let us set it aside for there are other more immediate problems for the reductive physicalist.

One such problem arises even if we were to grant the most liberal account of the existence of disjunction of states. The character of such a disjunction can be such that it precludes our regarding it as a physical state in which case it is not plausible to regard that disjunction as available to a physicalist reductionist. Consider again the belief that I will hear Lloyd better if I lean forward. There is a powerful attraction to Descartes's conviction that we might have existed as disembodied spirits deceived by an omnipotent demon. We may try to finesse this untutored conviction

---

* At least a part of Davidson's argument for anomalous monism is based on the possibility that the analogous question for mental states receives a negative answer. The issue of whether the "appropriate mapping" is one of the translations is significant but will be left dangling.

by distinguishing what is conceivable from what is possible and by asserting that although this is conceivable, it is not in fact possible. Still there is nothing in the way of our taking this untutored conviction at face value and saying in concert with Descartes, and the only mildly tutored masses (and David Lewis) that it is possible that I, this very person who is actually wholly physically constituted, be a purely or partly spiritual being. In that case, it might be that some of the closest worlds in which I believe that I will hear Lloyd better if I lean forward are worlds in which that mental state is constituted in part or wholly nonphysically. Then the disjunction my leaning forward counterfactually depends on *the actual world* has as part disjuncts referring to such possibilities. Lewis (1971, 209–210) seems committed to this possibility. He says, "Everybody is such that he might have been a disembodied spirit" is true, whereas "Everybody is such that it might have been a disembodied spirit" is false. The former means that each of those things that are both persons and bodies has a disembodied spirit as personal counterpart, whereas the latter means that each of the same things has a disembodied spirit as bodily counterpart. This does not mean that my belief in the actual world is less than wholly physically realized. We had granted this much right at the outset. The key thing to note here is that even granting the actual physical constitution of the world, we do not have a guarantee that the explanation of my leaning forward is wholly physical because an explanation is modally loaded, and what happens in other possible worlds is not uniquely determined by what happens in this world.

Lewis introduces his instructive example with the idea that the state of the lock, which is its being unlocked, is that state which has been responding in a certain way to being gently pulled as its typical consequence. This raises the question of whether the counterfactuals we assent to and indeed use to define the state, which is the lock's being unlocked, can be genuinely regarded as true according to Lewis. If there is no more to being a cause of the opening of the chain than standing in the appropriate counterfactual relation to that opening of the chain, then it seems hard to resist the idea that the alignment of the slots is a state that causes the opening of the chain. The only way of resisting that conclusion is to take Lewis's lead and deny that apart from $p\#$ that there is another state to be the alignment of the slots, but then the defining counterfactual is false. However, in that case, the opening of the chain, contrary to our initial uncontested intuition, does not depend counterfactually on the alignment of the slots. The threat then is that Lewis's approach, far from salvaging our ordinary commitments by way of an appropriate reduction, would lead to a wholesale repudiation of our ordinary conception of the world.

FUNCTIONALISM AND THE DENIAL OF THE IDENTITY THESIS

The argument I have given against the identity thesis is not another rerun of the celebrated multiple realization arguments of functionalists since Putnam have deployed against type identity theorists. For one thing, the argument I have given is not focused on the type identity but rather on the modal properties of particular tokens, which are not determined by their extensional characters. This shows that there are real and significant differences between the argument I have presented, which focuses on the way states and properties have a modal character and the functionalists' arguments that, for example, pain may be differently realized in different species or individuals or individuals at times. Indeed, as I shall endeavor to show, functionalism is itself liable to objections from this sort of argument.

When confronted by the functionalist's multiple realization arguments, it is open for the type identity theorist to relativize their identity thesis, much in the manner of Lewis (1980). In that case, the identity theorist may assert that they are interested in the reduction of pain in humans and not pain as such because pain as such may be realized in infinitely many differing manners. In that case, the relativized type identity theorist may venture to hope that there is indeed some physical property shared by all and only instances of pain in humans even if there is no physical property shared by all and only instances of pain as such. This sort of relativization will not work as a response to the modal argument I have presented because that argument focuses not on the type of belief but on the particular token belief and its modal properties. Any relativization then is beside the issue. The failure of the token identity theory will refute the relativized type theory.

It ought to be noted that I am not claiming that this failure is ubiquitous. Rather, my point is that as a general thesis, the token identity thesis is false: mental states cannot be one and all identified with the physical states that constitute them. The argument I have presented is consistent with the thought that there may be certain mental states, such as a pain or perhaps an itch, for which a token identity thesis or even a relativized type identity thesis is true. My claim is that this will not always be true and that more is needed to establish this sort of identity thesis than a mere commitment to the physical constitution of all.*

The modal argument against token identity theories cuts the ground out from under this sort of functionalism and also the relativized type identity theory because it denies the token identity of mental and physical states on which this functionalism depends.

Moreover, and importantly, the considerations presented thus far show that functionalism itself is similarly subject to worrying differences in counterfactual dependencies between the functionally characterized state, which is typically characterized in terms of counterfactuals or causation, and the realizing state, which is typically a state picked out in the language of a lower level theory. The question raised for a functionalist is why they should think that the so-called realizing state is indeed a state that realizes the modally characterized functional role. The argument presented has shown that if we use causal role as the criterion for characterizing the functional state and picking out the realizing state (which is characteristic of functionalism), then it is possible that there be a functional explanation without a physical realizer.† After all, mental states might nevertheless be defined in terms of the functional roles they play, although no physical states play just those roles.

---

* This holds even if we add to this thesis of physical constitution a further thesis, which I do endorse, of the supervenience of the mental on the physical.

† The reader will note that this argument also reveals a problem with Jerry Fodor's account of the relations among the various sciences, in his "Special Sciences" in *Representations* (MIT Press; Cambridge, 1981), because he too assumes that the theoretical explanatory autonomy of the special sciences requires not term-for-term reduction but something like a realizer for the terms of the special science in terms of "reducing" science.

Whether functionalism is false or not in this situation is a terminological matter. A functionalist account of the mind is a conjunction of two theses. The first is that at least some mental states are characterized by their functional role, spelt out by their characteristic causal roles. This thesis is untouched by the argument presented. The second thesis characteristic of functionalism is that there is a state not characterized in this manner, which is the realizer of that functionally characterized mental state. This can be taken as a constitution thesis, in which case functionalism survives, or it can be taken as an identity thesis, in which case the argument presented shows it can be false. Functionalists have not been too concerned about the distinction between these two ancillary theses. The argument presented here suggests they ought to show caution. The idea that physical reductionists can with impunity invoke functional roles and explanations and commit themselves only to the physical states that actually constitute the states is manifestly mistaken.*

Functional roles are characterized causally and, therefore, modally. Because a realizer is a state that plays a certain functionally specified (causal-explanatory) role, the actual constitution of the world will not suffice to ensure that the actual realizers of functionally characterized states are physical states, even if the actual world were wholly physically constituted. Because realizers of functional roles must support the explanatory inferences which, as we have seen, mere actual constitution does not support, the relations of constitution and realization are distinct.

EXPLANATIONS AND REDUCTIONS

In general, it follows that if we vigorously follow a reductive program, we can lose explanatory power. When we reduced the unlocked state of the chain to the state of its slots being aligned, we achieved an explanatory reduction. The state of the slots being aligned embeds just like the unlocked state of the chain. If we had gone further and identified the unlocked state of the chain with $p\#$, the state that constitutes the slots being aligned, we would have lost the explanation we had. For then, as we saw, the counterfactual dependence we needed for our explanations disappeared under analysis. This is an example of an important phenomenon: A pattern of dependence or causation can emerge at a higher level from a lower level that lacks such a pattern of dependence or causation. This is important for discussions of the way microlevel events are related to macrolevel events in general. This is the converse of an observation made by Lewis in his original paper on causation. There he noted that events may be related by causal dependence, which are not themselves directly related by counterfactual dependence because they can be linked by a chain of events, each of which linked by counterfactual dependence. This immediately leads to the thought that there can be counterfactual dependence of microstates without counterfactual dependence of macrostates. Here we see that the converse is also possible, that there can be counterfactual dependence of macrostates without counterfactual dependence of microstates.

Some explanatory counterfactual relations, such as, for example, that which subsists between my leaning forward and my belief that I will be able to hear Lloyd better if I lean forward, are at best captured as a physicalist mock-up in an infinitary

---

* For a clear statement of the view I have just denied, consult Pettit and Jackson (1988).

disjunction. Moreover, as we saw, as bad an account as this is, it is just not available if, as Lewis himself suggests, some relevant personal counterparts of mine are non-physical (Lewis, 1971). For in that case, the disjunction on which my leaning forward counterfactually depends involves disjuncts that can only be characterized nonphysically. My actual physical constitution does not necessitate that what is possible for me must be physical. The dependencies of my actual states are determined by what is possible for me. Thus, whether my leaning forward actually depends on my belief depends on how that belief might be constituted and what happens in some range of these possibilities. Insofar as our conception of the world is ineliminably causal, our conception of the world is ineliminably modal. The idea of a world characterized in purely extensional terms seems a misleading philosophical fiction.

At the outset, I asserted that this argument takes the form of a *reductio ad absurdum* of David Lewis's position. Of course there are many responses to any putative *reductio*. Let us be clear that if he had renounced his commitment to the "constitution is identity" thesis, this argument would have passed him by. However, in that case, he too would have denied the identity thesis. David Lewis did not do that as he was committed to an extensionalist conception of the world. I have tried to show that, contrary to Lewis's view, this extensionalism is inconsistent with our ordinary conception of the world. To take it seriously would be to repudiate much of our hard-earned knowledge of the world; in particular, it is in tension with our causal understanding of the world. Our commitments are modally loaded. The idea that the physical constitutes all gives one contentious but plausible version of the primacy of the physical. The primacy of the physical does not entail that all that is so constituted is physical, nor that all explanations are at bottom physical explanations.

We have seen that it is consistent with a commitment to the physical constitution of all to deny that explanations at the level of physics preserve explanatory relations available at other levels. The force of this conclusion is that it does not follow that when we regard something as physically constituted, the explanations we accept at a higher level will be replicated at a lower level. Explanations can be emergent. This is merely about what is possible. It does not show that explanations are inevitably emergent. The example of the lock shows us that reduction can occur. The question is how do things stand in biology? Can we be confident that whenever we have an explanation of evolutionary change at the level of individuals that this explanation will be replicated at other levels? The argument so far does not settle this at all. It does not tell us how explanations work in biology and whether they are reducible to lower level explanations. We now turn to that issue.

GENETIC DETERMINISM AND GENETIC REDUCTIONISM

The terms "genetic determinism" and "genetic reductionism" do not bear much resemblance to the theoretical positions of determinism and reductionism. Genetic determinism is the view that genes determine the phenotype. This needs to be distinguished from the view that genes cause the phenotype. The latter claim merely states that there is a causal process in which genes play a causal role that produces the phenotype. This much is clearly true. The claim of genetic determinism is that genes alone play a causal role. This is clearly false and no one really believes this view. Certainly, it is at odds with the usual understanding of the relationship between genotype, environment, and phenotype.

Genetic reductionism, however, is a distinct view, or rather a family of distinct views. It is characteristically presented by way of the theoretical claim that fitness, which Darwin thought was a property of individuals, should really be ascribed to genes and genes alone. By extension, we can define a measure that relates to individuals but that will be a weighted mean of the fitness of the individual's genes. The key theoretical purchase is afforded by the genetic level of analysis. Even here, there is room to wriggle. Is this a theory of accounting or a theory of explanation? Dawkins himself seems to have moved between these two positions. In his first book, *The Selfish Gene* (Dawkins, 1976), he argued vigorously for the explanatory priority of the genetic level of analysis. Later, he suggested that the genetic perspective is a useful perspective, not that it is the fundamental level of analysis.

In *The Selfish Gene*, Dawkins took himself to be articulating a perspective on evolution that had been maturing since the New Synthesis, the blending of Darwin's theory with the science of genetics. This was a gene-centric conception of evolution. If the New Synthesis had provided a story of heredity that got over Darwin's problem with the blending and eventual dissipation of variation, the new perspective saw the genetic story as providing a lot more than a mechanism of heredity. The idea was that fitness, which Darwin has associated with individuals, was really a property of genes. This idea was first formally proposed by the statistician R.A. Fisher in 1930. Fisher suggested that the proper object of evolution was the gene pool, the collection of interacting genes in a population. This proposal could be interpreted in two ways. The first was merely a heuristic. This perspective owed much to statisticians and a new breed of scientists, population geneticists.

One of the key theoretical problems debated in the years following the publication of *Animal Dispersal in Relation to Social Behavior* by V.C. Wynne-Edwards (1962) was the level of selection. What was it that was selected? Which features and what were the properties of these features? Wynne-Edwards suggested that sometime there are properties of groups that are selected for and which do not confer advantage to the individuals that possessed them. In this way, he argued for the existence of group selection; that is, he argued for adaptations that occur specifically at the group level rather than at the individual level. This claim was attacked by George C. Williams in a very influential book *Adaptation and Natural Selection*, a book that may stand for the neo-Darwinian orthodoxy. Williams (1966) saw no reason to grant that anything other than individual selection took place. It may be possible in principle but so vastly unlikely that it could be safely ignored. While Williams was articulating the orthodox response, orthodoxy received a jolt from the work of W.D. Hamilton, who had been working on kin selection and the related notion of inclusive fitness.

Dawkins popularized the tendency to focus on the genetic level, and in his book *The Selfish Gene*, he took what can be called the gene-centric perspective. In fact, Dawkins's view was a popularization of the work of G.C. Williams and W.D. Hamilton to a large extent. The language Dawkins deployed to explain the gene centric view were, as he acknowledged, metaphors and not to be taken literally. Genes were not literally selfish. Genes did not intend anything let alone their own survival through the bodies they constructed, and even that talk of constructing bodies has a strong metaphorical aspect to it. Genes act as if they had intentions, or so Dawkins says.

Dawkins distinguished between replicators and vehicles. This was a generalization of the distinction between genes and organisms. Genes are involved in the production of copies of themselves in transcription and also in the production of chains of amino acids in translation. For Dawkins, the key was that genes are involved in transcription; they replicate themselves. The idea he promoted was that all else that genes are involved in is shaped by the tendency to produce copies of themselves, that they are replicators. So what is a replicator in general? A replicator must have three important properties that Dawkins seemed to find uniquely but contingently exemplified by genes: they must last a relatively long time (longevity), they must be involved in their replication (fecundity) and they must have relatively speaking fidelity in that replication. For chromosomes to be replicators they would have to last longer than they do. Even though they are fecund and they last a relatively long time, crossing over means they do not have the kind of fidelity Dawkins thinks is required. If contingently they did have that fidelity then they too could be replicators.

Wilkins and Hull (2014) contend that Dawkins is not a genetic determinist but is a genetic reductionist. I want to suggest that this contrast is misplaced. There is only one position that is coherent to be a genetic reductionist because you are a genetic determinist. Without genetic determinism, there is no coherent position available to the genetic reductionist. So what is genetic determinism? This is the view that the phenotype can be regarded as functionally determined by the genes. So what is genetic reductionism? This is the view that the explanation for the various phenotypic traits observed is the differential survival of the genes that lead to them.

Two points to make before we go on. The first is that from the definition of replicator and vehicle, it is not clear that these are mutually exclusive. Could it be that some objects were both replicators and vehicles? Replicators are objects that get replicated, copied. Vehicles are also in fact copied and much in the same way that Dawkins's replicators are copied: by having similar copies made. A gene does not survive by itself by being present in descendants but by having copies of itself present in descendants.

Second, could we be genetic reductionists (of the sort Wilkins and Hull think of Dawkins as being) and therefore hold that the genetic level explains why phenotypic traits have the character that they do and also hold that there are other explanations for the phenotypic traits? That is, it is one thing to think that there is an explanation to be had in terms of survival of genes, and it is another thing to think that this is the only sort of explanation to be had. Dawkins does seem to be of the view that other sorts of explanation are ruled out by his account of replicators and the phenotypes they produce.

Genetic determinism is clearly false. The story of how we get to phenotypes is agreed on all sides to mention the role of the environment. This much is plain. Where disputes arise is in the significance of this universally accepted story. Dawkins suggests it is possible to hold to it and yet be a genetic reductionist. There seem to be very good reasons to think that this is mistaken. This seems to be at the heart of the dispute between Gould and Dawkins.

Having seen in the previous section on Lewis's identity theory of mind that we can have counterfactual dependence at one level that is not replicated at a lower level, we can readily understand that explanations can be emergent. The following would be such an example. Suppose we have a bunch of individuals with some phenotypic feature p that is adaptive and that feature comes to dominate in the population in the normal course of events. This is the sort of scenario Dawkins accepts as amenable to the genetic reductionist explanation. A genetic reductionist explanation would run something like this. These individuals share a genetic trait g, and the presence of g leads to p in the normal run of things. The survival of copies of g is what explains the survival of the phenotypic trait p. The explanatory direction runs from properties of the genetic level to the survival of the individuals with the genes.

Explanation is the crucial thing at this point. It is certainly true that part of the explanation for the presence of p is that the bearers of that phenotype have g. However, it is also true that part of the explanation for the presence of g is that it causes p the phenotypic trait that is adaptive.

So far, quite intentionally, the issue of what is being selected has been kept very vague. There was a justification for this in the proto-theory of natural selection. This theory pointed out that the theory of natural selection is really a theory of persistence. It seeks to explain why things are the way they are by adverting to their properties that help them persist. So what sorts of things persist? Anything heritable persists. In fact, heritability is the key aspect for Darwin. Were it not heritable, then a variation even if it helps its bearer to persist would not be present in the future.

Imagine a scientist who charts carefully the changes in frequencies of various gene variations in a population across space and time. These changes are the evolution of the population. In effect, this is the evidence accrued by population geneticists. However, this evidence, it is plain, is not the causal level at which we explain the changes but the effect of those causes. Just as watching shadows, we can discern facts about the objects that cast the shadows, so by observing the gene frequencies, we can figure out facts about the organisms that bear those genes. However, we would be mistaken were we to imagine that it is the shadows that bring about the changes. Similarly, there is a tendency to mistake the changes in gene frequency as the cause and not the effect of evolutionary changes.

One way to make that mistake is to think that it is the genes that determine everything significant about the organisms whose interactions is the interface of selection, as Dawkins puts it. Certainly, Richard Dawkins himself has a tendency to talk and think this way. His picture of selfish genes is just a figurative way of relaying that underlying causal story. It is because of the genes that individuals have the fitness that they have derivatively. This leads to the idea that in mapping genetic variations and their changes, we are mapping the most fundamental level of explanation in evolution.

Suppose that a genotypic trait—that is some property at the genetic level—is in fact a cause of a trait that causes the bearer of the trait to survive and reproduce. One consequence of that is that the genotypic trait occurs in later generations. Have we said enough to show that we can say that the selection occurred at the genotypic level? Many think we have. However, considerations show that the case has not been established. Suppose there is one phenotypic trait p that is a consequence of two distinct genetic traits g1 and g2. That is, within a range of environments we are considering, any individual with either g1 or g2 will have p. Now suppose that whether an individual survives to reproduce depends on having p. Certainly, the cause of an individual with g1 surviving is that it has p and that is caused in fact by g1. Does that show that that individual's survival depends on having g1? No, because it depends on having p and that depend on being either g1 or g2. This means that the individual's survival is dependent on being p and that is caused by being g1, but the individual's survival does not depend on being g1. The upshot of this realization is that an individual surviving to reproduce, which is acknowledged by everyone, even Dawkins, to be the interface of evolution by natural selection, can depend on having a phenotype that determines its adaptedness and that can be caused by a genotype but not depend on that genotype. Moreover, other combinations of genotype and environment could have led to the very same phenotype. What matters is that the interface of selection has the adaptive character required. Now notice that the significant point made by Dawkins—that the phenotype is extended—has the additional consequence that the level of selection is whatever the survival depends on and that can occur at many and distinct levels. A complication raised by this analysis is that to understand what was selected takes more than just knowing what causes led to the changes we see. It takes understanding what the persistence of different features depended on. This means being able to say how things would have gone in other nonactualized scenarios. This is quite a bit harder than what is hard in any case, spelling out what caused what.

REFERENCES

Anscombe, E. and P. T. Geach. 1954. *Descartes: Philosophical Writings*. Edinburgh: Thomas Nelson and Sons Ltd.

Cummins, R. 1975. Functional analysis. *Journal of Philosophy* 72: 741–765.

Darwin, C. 1859. *On the Origin of Species*. 1st ed. London: John Murray; 2nd ed., 1860; 3rd ed., 1861; 4th ed., 1866; 5th ed., 1869; 6th ed., 1872; 6th ed., 1876 with additions and corrections.

Dawkins, R. 1976. *The Selfish Gene*. Oxford: Oxford University Press.

Dawkins, R. 1982. *The Extended Phenotype*. Oxford: Oxford University Press.

Dawkins, R. 1984. Replicators and vehicles. In *Genes, Organisms, Populations: Controversies over the Units of Selection*, edited by R. N. Brandon and R. Burian. Cambridge MA: MIT Press. pp. 161–180.

Dawkins, R. 2004. Extended phenotype—But not too extended. A reply to Laland, Turner and Jablonka. *Biology and Philosophy* 19(3): 377–396.

Dennett, D. C. 1987. *The Intentional Stance*. MIT Press.

Dennett, D. C. 1991. Real patterns. *Journal of Philosophy* 88: 27–51.

Fine, K. 1975. Critical notice of counterfactuals. *Mind* 84: 451–458.

Fine, K. 1992. Aristotle on matter. *Mind* 101: 35–58.

Fisher, R. A. 1930. *The Genetical Theory of Natural Selection*. Oxford: Clarendon Press.

Godfrey-Smith, P. 1993. Functions: Consensus without unity. *Pacific Philosophical Quarterly* 74: 196–208.

Godfrey-Smith, P. 1994. A modern history theory of functions. *Noûs* 28: 344–362.

Gould, S. J. and R. C. Lewontin. 1979. The Spandrels of San Marco and the Panglossian Paradigm: A critique of the adaptationist programme. *Proceedings Royal Society London B* 205: 581–598.

Hajek, A. 2003. What conditional probability could not be. *Synthese*. 137: 273–323.

Hawks, J., K. Hunley, S. H. Lee, and M. Wolpoff. 2000. Population bottlenecks and Pleistocene human evolution. *Molecular Biology and Evolution* 17(1): 2–22.

Hume, D. 1748. An enquiry concerning human understanding. In *Enquiries Concerning Human Understanding and Concerning the Principles of Morals*, edited by L. A. Selby-Bigge. 3rd ed. Revised by P. H. Nidditch. Oxford: Clarendon Press; 1975.

Huxley, L., ed. 1901. *The Life and Letters of Thomas Henry Huxley*. Vol. 1. London: Macmillan

Huxley, L., ed. 1908. *The Life and Letters of Thomas Henry Huxley*. Vol. 2. London: Macmillan.

Johnston, M. 1992. Constitution is not identity. *Mind* 101: 89–106.

Kropotkin, P. 1902. *Mutual Aid: A Factor of Evolution.* 2009 paperback ed. London: Freedom Press.

Lewis, D. 1966. An argument for the identity theory. *Journal of Philosophy* 63: 17–25. Reprinted in D. Lewis, *Philosophical Papers*. Vol. I. Oxford: Oxford University Press, Oxford, 1983.

Lewis, D. 1971. Counterparts of persons and their bodies. *Journal of Philosophy* LXVIII: 203–211. Reprinted in D. Lewis, *Philosophical Papers*. Vol. I. Oxford: Oxford University Press, 1983.

Lewis, D. 1973a. Causation. *Journal of Philosophy* 70(17): 556–567.

Lewis, D. 1973b. *Counterfactuals*. Oxford: Blackwell Publishers. Reprinted with revisions, 1986.

Lewis, D. 1980. Mad pain and Martian pain. In *Readings in Philosophy of Psychology*. Vol. I, edited by N. Block. Cambridge: Harvard University Press. pp. 216–232. Reprinted with postscript in *Philosophical Papers*. Vol. 1. Oxford: Oxford University Press, 1983.

Lewis, D. 1986. Causal explanation. In *Philosophical Papers*. Vol. II, edited by D. Lewis. Oxford: Oxford University Press. pp. 214–240.

Michael, M. 2013. Problems with Lewis' argument for the identity theory. *Ratio* 26: 51–61.

Millikan, R. G. 1984. Language, Thought and Other Biological Categories. MIT Press: Cambridge, MA.

Neander, K. 1991. The teleological notion of "function." *Australasian Journal of Philosophy* 69: 454–468.

Patterson, B. 1956. *Early Cretaceous Mammals and the Evolution of Mammalian Molar Teeth.* Vol. 13, No. 1. Fieldiana: Geology. Chicago: Chicago Natural History Museum.

Peirce, C. S. 1955. The fixation of belief. In *Philosophical Writings of Peirce*, edited by J. Buchler. Dover Publishing.

Pettit, P. and F. Jackson. 1988. Functionalism and broad content. *Mind* XCVII: 381–400.

Stich, S. 1981. Dennett on intentional systems. *Philosophical Topics* 12: 39–62.

Wilkins, J. S. and Hull, D. 2014. Replication and Reproduction. *The Stanford Encyclopedia of Philosophy* (Spring 2014 Edition), edited by E. N. Zalta, http://plato.stanford.edu/archives/spr2014/entries/replication/accessed August 18, 2015.

Williams, G. C. 1966. *Adaptation and Natural Selection*. Princeton, N.J: Princeton University Press.

Wright, S. 1930. Review of the genetical theory of natural selection by R. A. Fisher. *Journal of Heredity* 21: 349–356.

Wright, S. 1931. Evolution in Mendelian populations. *Genetics* 16: 97–159.

Wright, S. 1960. Genetics and 20th century Darwinism. *American Journal of Genetics* 12: 365–372.

Wright, L. 1973. Functions. *Philosophical Review* 82: 139–168.

Wright, L. 1976. *Teleological Explanations*. Berkeley, CA: University of California Press.

Wright, S. 1980. Genic and organismic selection. *Evolution* 34: 825–843.

Wynne-Edwards, V. C. 1962. *Animal Dispersion in Relation to Social Behaviour*. Edinburgh: Oliver and Boyd.

6 Philip Henry Gosse and the Geological Knot

EXPERTISE AND THE OPENNESS OF SCIENTIFIC KNOWLEDGE

None of us have expertise in all the fields of knowledge on which we depend. We take our cars to mechanics, our children to physicians, and our cell phones to technicians, and we are often in no position to assess the claims these people make. We may be expert in one or two of these fields of knowledge but we are unlikely to be expert in all of them. This is not just a feature of our contemporary situation. This has been so for at least as long as we have historical records of our dealings with knowledge. The ancient Greeks distinguished between exoteric and esoteric knowledge. The former was open to all and so widely shared. The latter, on the other hand, was kept for initiates alone. The Hippocratic Oath, commonly attributed to Hippocrates of Kos, a fifth century BCE physician, has an intriguing clause:

> I will reverence my master who taught me the art equally with my parents, I will allow him things necessary for his support, and will consider his sons as brothers. I will teach them my art without reward or agreement; and I will impart all my acquirement, instructions, and whatever I know, to my master's children, as to my own; and likewise to all my pupils, who shall bind and tie themselves by a professional oath, *but to none else.*

Edelstein (1943)
Hippocrates, 56

Medical knowledge was esoteric and was to be kept secret. The discoveries, principles, tests, and treatments were to be kept within the inner circle whose members have taken the professional oath. This secrecy strikes us as perverse and indeed in tension with the essential democratic character of science. The idea that we would keep knowledge secret, and in particular that we would consider keeping it secret with a view to benefitting from that secrecy cuts against the basic conception of science as open to all. This openness of science has two aspects. First, that the body of scientific learning is a body of doctrine that is open to anyone who would seek to understand it, but also it has a second aspect, and this may be even more important: Any scientific claim can be challenged by anyone. It is not restricted to a professional class or guild or initiates into an esoteric cult. We all have the right to enter into the debates because the evidence and reasoning is open to all so that any theory can be an object of criticism.* This does not mean that just any comment on any theory is as significant a criticism of the theory as any other comment—democracy does not mean that all voices are equally compelling—but it does mean that the challenge "Why do you think that?" raised to someone propounding a scientific claim cannot be dismissed and must be answered.

It is in that context that the public distrust of evolutionary science can look very disturbing. However, it is almost reassuring to see that the distrust of science in general is high. There is a common motif that can be discerned in the rejections of public health policies like vaccination or fluoridation of water and the denials about climate change and evolutionary science. Thus, maybe there is nothing especially odd about evolutionary science? Actually, it seems evolutionary science is a special case and it is that because of the way it interacts with particular religious views.

RELIGION AND SCIENCE

There are two things to remember when we reflect on the relationship between evolutionary science and religion historically. First, many religious people have, from the first, been very ready to accept Darwin's theory as literally true. As Michael Ruse (2000) emphasizes, there are numerous tacks one might try to reconcile Christian faith with evolutionary theory. On the other hand, and second, it is also true that many religious people have felt that were Darwin's theory to be accepted as literally true, then they could not sustain the religious views to which they were actually committed. These are both witnessed and mark conflicting attitudes to the significance of Darwin's theory for religious belief. It is a further question whether people in the first category are all of one type. It might well be that some people who do not feel a tension between their religious commitments and Darwin's theory do not treat all aspects of their religious views with the same attitude. Perhaps some passages in their holy text are allegorical, or some passages are designed to give a glimpse of the deeper truth behind the words, or … indeed whether there are not among them people who are insensitive to the irrationality of their inconsistent commitments. For that matter, it might well be that there are people who felt a tension between Darwin's theory and their religious commitments who could be mistaken in thinking that there was such a tension. No one can be absolutely sure that they accurately discern the consequences of their views. In any case, the status of evolutionary science has been felt to have a serious import for many people and its status as science is a favorite target of challenge for those who feel it threatens dearly held religious convictions.

There is a tendency today to treat religion and science as though they are engaged in radically different endeavors. This is a very historically misleading view. Religion is not a priori in conflict with science, but it is *in fact* in conflict with science. This is not a new conflict either. The remarkable Derveni papyrus, the oldest European text we have, is an attempt to argue that the account of the Greek gods must be given an allegorical reading (Tsantsanoglou et al., 2006). It seems to be doing this to reconcile

---

* This tension is witnessed in the attitudes scientists take to the contemporary practice of keeping the results of many experiments secret on the basis of commercial-in-confidence. See, for example, Goldacre (2012).

the importance of the Greek gods culturally with the alternative explanations Greek philosophy (which included science) was developing. The suggestion if we were to generalize is that what we call religion is engaged in explaining the world, in the same way science is. Where there is a tension, there are strategies for separating the projects. The allegorical reading of traditional beliefs being the one the Derveni author is suggesting.

POPPER'S DOUBTS ABOUT DARWINISM

Is evolutionary theory a science? Even those who seem to be sympathetic to Darwin's theory, such as Karl Popper, have found themselves suspecting that its claims are not truly scientific. At one point, Popper (1976, 151) said "Darwinism is not testable scientific theory but a metaphysical research programme."

This pronouncement by one of the most eminent philosophers of science at the time led to understandable reaction. Popper had cause to reconsider his view, and later publications (Popper, 1985, 241–243; see also Popper, 1978) seemed to withdraw this claim, but not really:

> When speaking here of Darwinism, I shall speak always of today's theory—that is, Darwin's own theory of natural selection supported by the Mendelian theory of heredity, by the theory of the mutation and recombination of genes in a gene pool, and by the decoded genetic code. This is an immensely impressive and powerful theory. The claim that it completely explains evolution is of course a bold claim, and very far from being established. All scientific theories are conjectures, even those that have successfully passed many severe and varied tests. The Mendelian underpinning of modern Darwinism has been well tested, and so has the theory of evolution, which says that all terrestrial life has evolved from a few primitive unicellular organisms, possibly even from one single organism.
>
> However, Darwin's own most important contribution to the theory of evolution, his theory of natural selection, is difficult to test. There are some tests, even some experimental tests; and in some cases, such as the famous phenomenon known as "industrial melanism," we can observe natural selection happening under our very eyes, as it were. Nevertheless, really severe tests of the theory of natural selection are hard to come by, much more so than tests of otherwise comparable theories in physics or chemistry.
>
> The fact that the theory of natural selection is difficult to test has led some people, anti-Darwinists and even some great Darwinists, to claim that it is a tautology. A tautology like "All tables are tables" is not, of course, testable; nor has it any explanatory power. It is therefore most surprising to hear that some of the greatest contemporary Darwinists themselves formulate the theory in such a way that it amounts to the tautology that those organisms that leave most offspring leave most offspring. C.H. Waddington says somewhere (and he defends this view in other places) that "Natural selection … turns out … to be a tautology." However, he attributes at the same place to the theory an "enormous power … of explanation." Because the explanatory power of a tautology is obviously zero, something must be wrong here.
>
> Yet similar passages can be found in the works of such great Darwinists as Ronald Fisher, J.B.S. Haldane, and George Gaylord Simpson; and others.
>
> I mention this problem because I too belong among the culprits. Influenced by what these authorities say, I have in the past described the theory as "almost tautological," and I have tried to explain how the theory of natural selection could be untestable (as

is a tautology) and yet of great scientific interest. My solution was that the doctrine of natural selection is a most successful metaphysical research programme. It raises detailed problems in many fields, and it tells us what we would expect of an acceptable solution of these problems.

I still believe that natural selection works in this way as a research programme. Nevertheless, I have changed my mind about the testability and the logical status of the theory of natural selection; and I am glad to have an opportunity to make a recantation. My recantation may, I hope, contribute a little to the understanding of the status of natural selection.

Therefore, unhappily, Popper's recantation in these comments amounts to his claim that natural selection is a successful metaphysical research program. He seems still to have accepted that Waddington and others were correct to say that it is a tautology and so not testable. The suggestion is that the theory is, however, useful in science for the same reason that logic and mathematics are useful in science. They are not testable but empty tautologies and so applicable generally. This is the view that we saw gutted the theory of natural selection in Chapters 2 and 3. We saw how this claim just does not give us a theory we can use to explain the world. Following this suggestion, we would be left with a theory that does not distinguish between the claim that something evolved and the suggestion that it evolved by natural selection. That distinction we have seen of paramount importance because the second gives us a mechanism for evolutionary change, something left open by the first.

It is important to see what conception of science might lead to the idea that evolutionary theory is not science and to see that it is that conception of science, that is, to clarify this issue, we have to be clear about the nature of science and the nature of evolutionary theory. Thus far, in this book, we have focused on the nature of evolutionary theory. We have seen that it is a theory that is aimed at explaining why certain varieties survived and others tended not to. It does that by identifying features that cause a differential rate of survival. This is all, by its very nature, historical. It can be likened to reverse engineering. We have the results and we seek the steps that led to it, *the why* as Aristotle put it.

The question whether evolutionary theory is science or not was made salient by the endeavor of the Logical Positivists and after them by Karl Popper to demarcate science from nonscience. For the Positivists, this was an important project because science and science alone was the embodiment of rationally acquired commitments. For Popper, this was not quite the only form of rationally acquired commitments but he was very concerned, as were the Positivists, with the unmasking of pseudoscience that masqueraded as science. Psychoanalysis and Marxism, for Popper, were frauds that draped themselves in the garb of science but failed in crucial ways to meet the criteria of science. That there are such criteria for science and whether it can be demarcated from nonscience is itself a far-from-settled matter. Philosophers, sociologists, historians, and scientists themselves have had a go at making sense of the demarcation question. One thing to say is that, like most concepts we deploy, we understand the nature of science by looking at paradigm exemplars and, like most real-world cases, in the case of science, we should not really expect a clear boundary between science and nonscience. In many areas, vagueness is real and not to be avoided. Consider for example, the perfectly well-understood notion of an adult. There are clear cases of adults and clear cases of juveniles but there are also cases that are vague. There is no sharp boundary between the adults and the nonadults. So it is with many of the concepts we deploy, and the concept of science is one of them. When we do consider the paradigm examples of science, science well exemplified, what looks to be a distinctive character seems to be something like a vulnerability of the theorizing to empirical checking. That, at the very least, is the self-conception of science. Some might deny this is at all possible; some might suggest that this is an ideal that scientists rarely if ever measure up to; some might suggest that this talk of empirical checking is mere ideology serving to coat the claims scientists make with a patina of respectability. We cannot sidestep this question. If we are going to assess the status of the claims of evolutionary science, then we have to face these questions about science in general and then see how things stand with the empirical vulnerability of evolutionary science.

The first order of business then is to get clear about the empirical nature of theorizing that seems important within science. Is this the reason to pay attention to scientific claims, that these are claims that we know to be empirical and which have, for that reason, some sort of justification? The way we treat scientific theories does seem to suggest that these claims have a special status. However, for that to be the case, two things need to hold. First, that story about the empirical aspect of theories needs to make sense, and some have said it does not. Second, assuming the general story about empirical vulnerability makes sense, the situation with evolutionary science also has to be appropriate with respect to empirical vulnerability.

First let us deal with the general claim. Is empirical vulnerability even possible? Suppose the denialists about science in general were correct. Suppose that there is nothing particularly rational about scientific claims. Then, there really would be no more reason to believe the account scientists give about the origin of humans than the account in Genesis. That is the issue we really want to face. Is the evolutionary story any more rationally compelling than the Genesis account? We will look at this question by first asking whether there is anything rationally compelling about science in general and then ask whether evolutionary theory is scientific. The first conclusion we are heading toward is that science is indeed different in character from religion. The second question about the status of evolutionary theory is not so straightforward. There are subtleties to negotiate.

RECONCILIATION BY DISPLACEMENT

A bit of historical context helps understand how we got to the present situation. Even before Darwin's *Origin* exploded onto the scientific scene, changing all that had come before, geologists had laid the foundation for his reconceptualization of the biological world by developing an account of the world that emphasized the transformative power of mechanisms contemporarily visible. Uniformationists insisted on the role of earthquakes, volcanoes, glaciers, flooding, sedimentation, and erosion to explain all the geological phenomena. This doctrine had its origin in the unlikely amateur scientist James Hutton (1726–1797), a Scottish farmer. Hutton realized that for these mundane mechanisms to play the role of explaining the way the world is, the Earth had to be significantly older than had been determined Biblically. In the seventeenth century, Irish Archbishop of Armagh, Bishop James Ussher had determined the date of God's creation by adding up the ages of the biblical figures right back to Adam and his erstwhile helpmate, Eve. With this textual method, Ussher determined the date of Creation remarkably precisely (if not accurately) as dusk on October 22, 4004 BC.

The Uniformationists' threat to the Genesis story predated Darwin and laid the scene for one of the more curious chapters of the relationship between religion and science: Philip Henry Gosse was a fascinating man and, luckily for us, the father of Edmund Gosse the first professor of English Literature at Oxford University (Figure 6.1). Gosse the younger wrote the endearing *Father and Son* (Gosse, 1907), an account of growing up with his deeply religious parents. A sensitive and friendly man, Gosse the elder was a fundamentalist religious man and a scientist of such note that he had received word from Darwin about his developing but as yet unpublished theory that would explain the observations of fossils, of shared characteristics among diverse organisms, of vestigial characters, etc., by dint of common descent. This shocked Gosse. As a fundamentalist, he believed in the literal truth of the Bible. Yet after considering the evidence Darwin had pulled together—its variety, its simplicity—he saw that Darwin had developed a strong argument for the idea of the mutability of species and that Darwin had identified a mechanism that would explain the variety and adaptation observed in the natural world. How could Gosse resolve this tension? He wrote a book, *Omphalos*. The name is the Greek term for navel and the title was chosen by Philip Henry Gosse very intentionally (Figure 6.2).

1857

FIGURE 6.1   Philip Henry Gosse and his son Edmund. (From Gosse, E., "Father and Son: A study of two temperaments," W. Heinemann, London, 1912.)

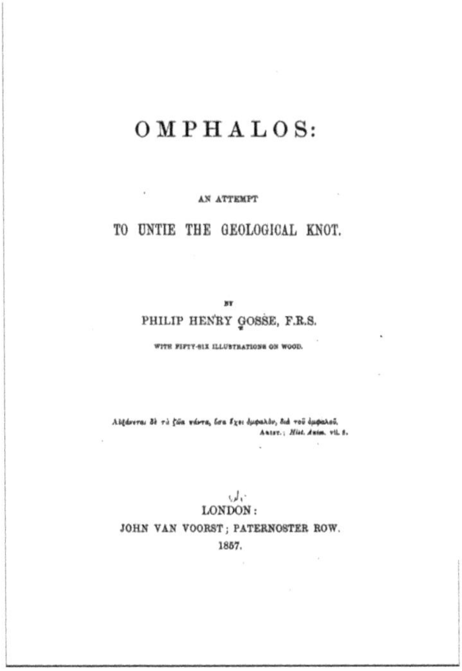

OMPHALOS:

AN ATTEMPT

TO UNTIE THE GEOLOGICAL KNOT.

BY

PHILIP HENRY GOSSE, F.R.S.

WITH FIFTY-SIX ILLUSTRATIONS ON WOOD.

Αὐξάνονται δὲ τὰ ζῶα πάντα, ὅσα ἔχει ὀμφαλὸν, διὰ τοῦ ὀμφαλοῦ.
ARIST.; *Hist. Anim.* vii. 8.

LONDON:
JOHN VAN VOORST; PATERNOSTER ROW.
1857.

FIGURE 6.2 *Omphalos*—coverpage.

The highly intelligent and fair-minded Gosse could see that, contrary to what many of his co-religionists who felt threatened by Darwin's arguments would say, the evidence Darwin had collected was indeed powerful. However, Gosse who was an utter literalist about the Biblical account of the world also thought he could see that there was a gap between the evidence and the theory. He wanted to show how one could agree with Darwin about the evidence gathered and yet disagree with the conclusion Darwin drew from that evidence. In suggesting that there is a gap between the evidence and the theory, he was correct. Whether the gap he identified really did allow for the rational reconciliation of his religious convictions with his scientific scruples is another matter we will examine later.

> I remember, when I was in Newfoundland, some five-and-twenty years ago, the disastrous wreck of the brig Elizabeth, which belonged to the firm in which I was a clerk. The master had made a good observation the day before, which had determined his latitude some miles north of Cape St. Francis. A thick fog coming on, he sailed boldly on by compass, knowing that, according to his latitude, he could well weather that promontory. But lo! about midnight the ship plunged right against the cliffs of Ferryland, thirty miles to the south, crushing in her bows to the windlass; and presently went down, the crew barely saving their lives. The captain had not allowed for the polar current, which was setting, like a sluice, to the southward, between the Grand Bank and the land. (Gosse, 1857, iii–iv)

The captain had tried to keep the ship out of harm's way. However, the captain's calculations would have been appropriate to keep the ship safe only on the assumption that the current made no discernible difference to the progress made by the ship. Such an assumption—a *ceteris paribus* or all else being equal assumption—is not always appropriate: sometimes things are not equal. Sadly, the captain lost his ship on the cliffs of Ferryland. Gosse also illustrated the role of such assumptions from mundane cases, an archer missing the bull's-eye because he had not allowed for the wind, to important scientific cases, such as the perturbations in the orbit of Uranus, which elicited many and varied responses from astronomers. The observations did not fit with the Newtonian law of gravitation. Some blamed the observations. The law (they held) was sound, so there must have been a mistake with the observations. Others held that the observations were correct and that the law of gravitation, which works so accurately on Earth and nearby the Earth, must be limited, not applying in full generality across the universe. Of course, as Gosse points out, both had ignored that there was another possibility: that there could be another body in the vicinity whose gravitational force was at work here, as it happened, Neptune.

> In each of these cases, the conclusions were legitimately deduced from the recognised premises. (The archer's) skilled eye had calculated the distance; his experience had taught him the requisite angle at which to shoot, the exact force necessary, and every other element proper to insure the desired result, except one. There was an element which he had overlooked; and it spoiled his calculations. He had forgotten the wind. (Gosse, 1857, v)

By analogy with these cases, Gosse wanted to argue that geologists who have argued for the antiquity of the earth have made sound calculations but their conclusions lead to error because "they have not allowed for the Law of Prochronism in Creation" (p. vi). This Law of Prochronism was one he himself had fashioned. Prochronism is the way any stage in the life cycle of an organism points to other stages in the cycle. The seed points to the flower whose fertilization led to it. The flower points to the adult plant that bears it. The adult plant points to the seedling that grew into the adult plant. The seedling points to the seed that grew into the seedling. So any point on the cycle points to every other point.

> The cow that peacefully ruminates under the grateful shadow of yonder spreading beech, was, a year or two ago, a gamesome heifer with budding horns. The year before, she was a bleating calf, which again had been a breathless foetus wrapped up in the womb of its mother. Earlier still it had been an unformed embryo; and yet earlier, an embryonic vesicle, a microscopically minute cell, formed out of one of the component cells of a still earlier structure—the germinal vesicle of a fecundated ovum. But this ovum, which is the remotest point to which we can trace the history of our cow as an individual, was, before it assumed a distinct individuality, an undistinguishable constituent of a viscus—the ovary—of another cow, an essential part of her structure, a portion of the tissues of her body, to be traced back, therefore, through all the stages which I have enumerated above, to the tissues of another parent cow, thence to those of a former, and so on, through a vista of receding cows, as long as you choose to follow it.
>
> This, then, is the order of all organic nature. When once we are in any portion of the course, we find ourselves running in a circular groove, as endless as the course of a blind horse in a mill. It is evident that there is no one point in the history of any single creature that is a legitimate beginning of existence. Additionally, this is not the

law of some particular species but of all: it pervades all classes of animals, all classes of plants, from the queenly palm down to the protococcus, from the monad up to man: the life of every organic being is whirling in a ceaseless circle, to which one knows not how to assign any commencement—I will not say any certain or even probable, but any possible, commencement. (Gosse, 1857, 121–122)

The cow is as inevitable a sequence of the embryo, as the embryo is of the cow. Looking only at nature, or looking at it only with the lights of experience and reason, I see not how it is possible to avoid one of these two theories. (Gosse, 1857, 122–123)

The circle of life has no natural beginning. Every point on the circle points to an earlier point on the circle. The embryo depends on the cow as the cow depends on the prior existence of an embryo. The circle was as well founded as any empirical theory could be (Figure 6.3). This was the dilemma: if God were to create the world at any point with an element of the circle in it, that cow or calf or embryo must be created at some point in the circle. In that case, let's suppose He chooses to create a cow. The cow as created would point to earlier moments on the circle, moments that on the supposition that God chose to create the adult cow simply did not exist. Similarly, were He to have created a calf, such a calf would point to a prior embryo and the cow that carried it. Not even God could create a cow without it pointing to the calf.

Those unreal developments whose apparent results are seen in the organism at the moment of its creation, I will call prochronic, because time was not an element in them; while those which have subsisted since creation, and which have had actual existence, I will distinguish as diachronic, as occurring during time. (Gosse, 1857, 124–124)

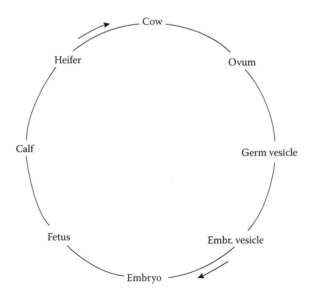

FIGURE 6.3 Cow circle from Gosse's *Omphalos.*

As Gosse argues, if God were to create the world, then He would have had to create it at a certain point in time. There is no other way to do it. However, whichever point of time He selected, at that point, the created world would have contained cows at some point of their development. Trees would have been created at a certain size with tree rings indicating that they had lived for a certain time. Similarly, when God created Adam, he would have been created with hair and nails, although hair grows from follicles on the scalp and nails grow from the cuticle at the base of the nails. In so creating Adam, God would have created a fully formed adult *as though he had had a history.* In particular, he would have created him as though he had lived long enough to grow hair and nails. If we think about it for a moment, we see even the idea that God creates Adam as an adult and humans are not born as adults means that Adam is presented *as if* he had grown from a newborn baby to his created adult form. So Gosse suggests, God must have decided in His wisdom to create each thing in His creation which has prochronic aspects, that is, each thing points to unreal *as if* elements in their *as if* past. Gosse suggests this is so for the world as a whole. There will be aspects of the world that suggest they had been caused, for example, by the slow wearing of mountains for millennia. This too is an *as if* history of the world.

The science of the past as practiced by geologists and by Darwin influenced biologists, as Gosse saw it, is the cataloguing of this *as if* history of the world. A botanist looking at created trees would correctly call one a juvenile and one an adult and even be able to assign them ages on the basis of tree rings. However, being created by God at the same time, they would in fact have exactly the same real age. The tree rings only indicate the *as if* age of the trees. In much the same way, a physician confronted with the adult Adam could have talked about the way Adam's hair and nails, and body as a whole, grew. This physician is accurately charting the *as if* history of the Adam portion of the world but not the *actual* history of the world.

To many who encountered his attempted rapprochement, Gosse's suggestion was literally incredible. How could they believe this strange combination of views? The idea that the world was created at some particular point in the past with an "as if" history depicted in it did not achieve the rapprochement he sought between his fellow biblical literalists and the scientists he also counted himself among. It needs to be remembered that Gosse was no mediocre scientist. He was a Fellow of the Royal Society and a scientific correspondent of Charles Darwin. Darwin himself sought his advice on "undersea battles" between organisms, for example. Gosse's fellow biblical literalists did not appreciate his conceding the point to the evolutionists that the evidence seemed to point to a long history. Conceding this to the scientists would mean that a special case had to be made, that contrary to all the evidence, the world had a beginning at a particular time, a special case that depended on a literalism about the Bible. That was one problem but there was another theological problem Gosse faced; it seemed that accepting Gosse's conciliatory cluster of views forced believers to accept that God must be seen as a deceiver. Gosse was conceding something the literalists were reluctant to countenance: that the evidence pointed to a long history for the Earth, God put the evidence there, and it is misleading.

His fellow scientists were no more inclined to accept Gosse's pacific efforts than his Biblical literalist brethren. The idea that the science of the past was not describing the history of the world but only an account of a fictive "as if" past turned that science into something aimed not at truth but something much less significant in their eyes, a game of what is true in a certain—never realized—fiction. To scientists, this gave up on the idea that science, which aimed at truth, followed the evidence. The evidence points to a long history but the truth is quite different.

In making his suggestion, Gosse was in fact replaying Cardinal Robert Bellarmine's challenge to Galileo, a challenge Descartes had tried to answer in his *Meditations*. In effect, when Galileo pointed to the evidence and suggested that his theories best explained the evidence, Bellarmine asked him to do more than just explain the appearances on the assumption of the heliocentric theory.

> I say that if there were a true demonstration that the sun was in the center of the universe and the earth in the third sphere, and that the sun did not travel around the earth but the earth circled the sun, then it would be necessary to proceed with great caution in explaining the passages of Scripture which seemed contrary, and we would rather have to say that we did not understand them than to say that something was false which has been demonstrated. But I do not believe that there is any such demonstration; none has been shown to me. *It is not the same thing to show that the appearances are saved by assuming that the sun really is in the center and the earth in the heavens.* I believe that the first demonstration might exist, but I have grave doubts about the second, and in a case of doubt, one may not depart from the Scriptures as explained by the holy Fathers. (Cardinal Robert Bellarmine, Letter to Foscarini on Galileo's theories, April 12, 1615)

He said, and he is right about it, that showing that the heliocentric assumption preserves the appearances is not the same thing as demonstrating the heliocentric theory. Demonstration, as he and his contemporaries used the term, is providing certain proof, not just balance of probabilities. The picture of demonstration comes from Aristotle who thought of demonstration as a derivation from first principles known by intuition. It is hard to find this picture other than implausible as an account of science. The standard it sets, certainty, is not within our ken. Because Einstein's theory replaced Newton's, a theory that had been incredibly successful for centuries, we all have taken on board that our best scientific theories are not certain but only provisional. That provisionality is not a fault with science but a sign of its openness to criticism. The requirement that we be able to prove our theories true is a requirement that comes from a conception of scientific inquiry at odds with our experience. This feature of provisionality of our scientific commitments is what makes the demand for vulnerability to the world, the sort of thing Popper regimented into his falsificationism, plausible and desirable. Bellarmine's challenge then is to prove that following the evidence leads to the truth. This is a maddening challenge to those of us with a scientific or philosophical bent. We have accepted that it is rational to follow the evidence and the idea that the evidence furnishes us with reasons to believe. However, there is a tradition that evidence must be contrasted with what is believed to be true. This can be seen among religious thinkers but it is exemplified among scientists too and it has a very long history. From Sextus Empiricus (*Outlines of Pyrrhonism* 1.33) we learn of Anaxagoras' example of this mode of argument: "Anaxagoras set the appearance that snow is white in opposition to the claim that snow is frozen water, water is black, therefore snow is black" (Curd, 2007, 121). The Greeks saw Anaxagoras' claim here as paradoxical but understood it as showing the rationality of assessing what appears to be evidence: if we know that water is black, then the evidence that it is not black but white is misleading. As Sextus puts it, "We set what is thought (i.e., antecedently committed to) in opposition to what appears." (1.33) Similarly, if we believe the universe is a series of crystal spheres centered on the Earth, then any evidence that, for example, there are moons around Saturn that seem to smash through the sphere on which Saturn resides must be misleading and can be ignored. This is not an attitude that disappeared with the scientific revolution in the early modern period. When asked what he would have thought if Eddington's famous 1919 gravitational lensing measurements had gone against his theory, Einstein is supposed to have remarked that he would have felt sorry for God because the theory was correct. Even allowing for the sort of rhetorical flourish Einstein was known for, this is not far from Anaxagoras' attitude.

What do we say about Gosse's supposition? It is quite tempting to put it aside as unrealistic or fanciful. Perhaps for those reasons it has long been lampooned but thinking about it is really quite worthwhile because it shows us the conflict between a reason-centered approach and one that gives a special role to nonempirical revelation. Some decades after Gosse, Bertrand Russell considered the possibility that the world as a whole was created five minutes ago:

> There is no logical impossibility in the hypothesis that the world sprang into being five minutes ago, exactly as it then was, with a population that "remembered" a wholly unreal past. There is no logically necessary connection between events at different times; therefore, nothing that is happening now or will happen in the future can disprove the hypothesis that the world began five minutes ago. Hence, the occurrences which are called knowledge of the past are logically independent of the past; they are wholly analysable into present contents, which might, theoretically, be just what they are even if no past had existed. (Russell, 1922, 159)

Russell was inclined to treat this hypothesis as a logically tenable but uninteresting skeptical hypothesis. For him, it shared the characteristic of all skeptical hypotheses: they were unlivable, reflecting David Hume's influence on Russell. Skeptical hypotheses would lead to inertia and yet we find ourselves having to act in the world and acting on beliefs that are in conflict with the skeptical hypotheses. Russell never felt satisfied that he had found a way to reconcile those facts. Hume held to a kind of naturalism: it is our nature to believe what we perceive so that skeptical hypotheses do not get a look in. Hume thought that as we consider matters of theory, of high philosophy, or the underlying causes of manifest phenomena for example, there the skeptical hypotheses gain most traction and their undermining force leads to a skeptical ambivalence on all such matters. The naturalism Hume advocates, that beliefs reflect our sensitive and not our cogitative nature, was not attractive to the mathematician Russell.

Gosse, though, thought he had a convincing reason to propose his *Omphalos* hypothesis: he was already committed to the literal truth of the Genesis story. Were that story true, then evidence that seems to point to theories inconsistent with it must be false. What are the lessons we can learn from Gosse's efforts? What light does this sort of story have to shed on our relationship to claims about the past?

The first point to note is that in trying to discover what happened in the past from the evidence we have, there are inferences to the best explanation that Gosse acknowledged were (taking into account their "oversight" of the literal revealed truth of Genesis) plausible supports for the age of the Earth far exceeding the few millennia allowed by Genesis. That is, the suggestion that the world had a long history has the same sort of plausibility as would the tree rings in a newly created world has existed for as many years as there are tree rings.

Evidence of this sort is always going to require inference to the best explanation. One view of this situation is say that this just goes to show that evolutionary biology (and presumably any attempt to find out about the past) is not genuinely a science. Perhaps physics is science but it might turn out that not much else will be; in the disrespectful bon mot attributed to the physicist Ernest Rutherford, "there's physics and everything else is stamp-collecting." In turn, this sort of attitude has itself come to be denigrated as "physics envy."

What should we say about it? Science is the rational inquiry into the nature of the world. We should not be too quick to think that we are already clear about either the nature of the world or the nature of that rational inquiry into the world.

All the evidence we can turn to about the past is present evidence. We have no direct access to the past facts but only through their present traces. Fossils, traces of past organisms that we discover, are present traces. Photographs, electronic records, books, memories, and so on to be evidence must be present to us. An account of a fossil once found but now lost is a present account. Images are present objects. Books with accounts of the past are present objects. Memories of past discussions are present memories. In every instance, the evidence we depend on is present evidence. The past is not however present. Therefore, we need to make an inference, not logically guaranteed, from the present evidence we have to the character of the past. We might have a present text or photograph that we take to show that something was the case in the past. We take the present photograph, for example, to be a causal consequence of a past event that we infer to have certain characteristics that, taken together, explain the features of the photograph. Such inferences are part and parcel of making sense of the past and in so doing making sense of the present traces we have of the past. There are also present absences of traces that are themselves in need of explanation. Why don't we have evidence of soft tissue of many mammals from the Cretaceous? This too needs explaining. In this case, we think the explanation is ready to hand: the rapidity of decay of soft tissue relative to harder materials, we think, explains the relative scarcity of soft tissue fossils from the Cretaceous mammals.

The study of the history of our world, in all its facets, depends on discerning present features of the world and trying to show how these features can be regarded as traces of the past. Fossils presently existing can be regarded as traces of past organisms. One example of this, from a very different field, is the very low energy background radiation discovered to fill the universe, the so-called 3 Degree Kelvin Background Radiation. This observation has been treated as the strongest evidence for the hot big bang and is surmised to be the trace of that big bang. Notice that, in this case too, the evidence can be observed presently is independent of the historical story into which it is fitted. This is also true about scars as evidence of past wounds. Of course, one rather important example of such a scar is the navel (or The Omphalos as discussed earlier).

Present memories can be regarded as traces of past events, and so on. The curious thing about memories is that they do not have this separation of present evidence from the past cause. The way memories strike us is as caused by such and such events in the past, and this phenomenology persists even when there is evidence that this is illusory.

Having a navel does not logically imply having had an umbilical cord. It certainly normally is caused by having had an umbilical cord and having had that cord removed but there is no necessity connecting having a navel with that normal course of events. When we try to find out about the past, whether the recent social past or the evolution of some organism or the formation of a geological structure or the past at a cosmological level, we can only depend on present evidence. For the social past, we rely on present memories, present photographs and writings, and so on. For the biological history of the world, we look to how things are now and consider how the environment has changed, how the organisms have changed in a manner that allows us to find explanations of what we see as the present situation. We move from how things are to an account that explains why they are that way. Is no such activity truly scientific? That is the sort of view one would associate with Eddington's dismissal of all such activities. As the example of the cosmic background radiation shows, this sort of dismissal would not just affect biology and geology but physics itself or, at least, cosmology.

There is a difference between the historical sciences and those which make clear predictions. The ones that make clear predictions fit a simpler predictive/experimental model. The term *science* is not really worth fighting for. That is merely a terminological dispute. The key point worth worrying about is that by drawing the line around what counts as science too restrictively, we end up misunderstanding the extent to which we can find out about the world rationally. Even when we do find things out rationally, this does not mean that we can be certain that we know, nor does it mean that we can show our view is better than all comers. It nevertheless is what we need to aim at. This is why the discussion of Gosse is important. He is highlighting the fact that there are auxiliary hypotheses involved in the historical explanations we seek. Someone might, as Gosse actually does, ask why their auxiliary hypotheses are not legitimate when everyone needs to make such hypotheses. Gosse assumes that God created the world at some point with adult cows and calves popping into existence. Even assuming that he has some sort of convincing answer, there is a crucial asymmetry between the auxiliary hypotheses Gosse invokes and those made by the evolutionary scientist. Gosse himself accepted that there is an *as if* history correctly described in Darwin's *Origin*. The evidence available to everyone points to the long duration that organisms have been about and have evolved from a common ancestor. However, absent a reason to think that the past was relevantly different from the present, can we make auxiliary hypotheses of the sort Gosse deploys?

IS EVOLUTIONARY THEORY SCIENTIFIC?

When I was an undergraduate student of biology, I first came across those who doubted that evolution took place and who doubted the scientific status of the claims scientists made about the evolution. At the time, I thought it would be a simple matter to put them straight. Just line up the evidence, it seemed to me, and they would see why they too should accept Darwin's theory of natural selection and the theory of evolution. Doing that did not seem to work, but just why it did not work was not clear. Was the doubt about the fact that evolution had taken place? That seemed very implausible. The evidence that showed there had been organic change over time was overwhelming. Was it that we had not clarified what the theory of natural selection was and what it explained? Was it therefore about the mechanism of evolution? Although the ostensive target was the power of the theory of natural selection, but the real target seemed to be the reality of evolution as such. What stuck in my mind was that Darwin's mechanism was the focus of the attacks but that even if those attacks on natural selection were successful, that would go no way toward showing that evolution had not taken place. In fact, there had been scientific debates about the mechanisms of evolution. The issue of whether Darwin's mechanism was indeed the main driver of evolution was a serious question, but none of these debates contested the reality or the evidence for, evolution itself.

The nature of the disputes over natural selection has to be seen in this context (Ruse, 2006). We have to separate the questions of the reality of evolution and the evidence for that from the questions about mechanisms that drive evolution. Even if, as I believe we would not, we decide that natural selection is unscientific or powerless to explain the phenomena Darwin sought to explain by reference to it, this would not show that evolution did not take place. There are questions about the nature of natural selection and whether the claims we make in deploying the theory of natural selection are—properly speaking—scientific. Even here, it is not so straightforward. Is it clear what it means to say that it is scientific? Is it the same question as asking whether a theory is a good scientific theory? In the following section, we turn our attention to this matter.

THE POSITIVIST STORY: INDUCTIVE LOGIC AND CONFIRMATION

The early twentieth century saw a revolution in many of the most fundamental scientific ideas. The progress was in some measure attributable to the way scientists came to rethink their approach to fundamental concepts and in particular to think about the way in which measurement (what we might call "observation") was connected to the theories we develop. Einstein's important influence was exemplified in his operationalizing of the concept of simultaneity. Einstein took a concept everyone thought they understood clearly and unambiguously—the idea of two events occurring at the same time—and proposed an empirical test for simultaneity. This might not seem very significant but operationalizing the concept turned out to lead to the possibility that whether events are simultaneous depends on the perspective from which they are perceived, that is, where and when the test is run. Again, intuitively, it seems that there is an objective question of whether two events are simultaneous but once the concept was operationalized, possibilities became conceivable that were not beforehand. Then, the possibility arose that two events could be simultaneous relative to one frame of reference and not simultaneous relative to another frame of reference. This relativity of simultaneity was a startling theoretical insight. That idea, take a concept which we think we understand and sharpen it by operationalizing it, by providing an empirical test for the application of that concept, drove much early philosophical thinking about how to do science and is easily seen to be a motivation even today. Often, it is a useful intellectual exercise. It asks the scientist to specify conditions under which they would accept that a concept is to be applied and conditions under which they would reject the application of the concept. Useful concepts would seem to be those that clear conditions of application. Or, at least, that is the ideal.

THE PARADOX OF CONFIRMATION

In the 1920s and 1930s, although the Nazis and their intellectual kin found solace and inspiration in the Romantics with their emphasis on intuition, manifest destinies, and race theories, there were other philosophers who instead found inspiration in the Enlightenment with its emphasis on the universality of reason and the importance of education. These were called the Logical Empiricists or Logical Positivists. They took as their task the reordering of society in accordance with reason and the model they took of reason was science. According to their understanding, science, being a rational activity, seeks to base its theories on evidence. They tried hard to understand just how theories are based on evidence. This turned out to be a trickier problem than it might have seemed. They took as one of their intellectual challenges spelling out just what it means for a theory to be supported and to what degree by some observations. The question is Given a body of observations, which theory should we believe? They also saw the role of the philosopher of science to be the production of a handbook for scientists, to guide their theory development and theory choice.

Deductive logic is, as Kant famously emphasized, nonampliative. Deductive logic does not amplify or extend our knowledge; it merely makes explicit what we had already committed to implicitly. Given some premises, the only things that follow deductively from the premises are already implicitly in the premises. For that reason, deductive logic seems to have little to do with the advance of science. The Logical Positivists thought what was needed was the development of a type of logic that went beyond mere deductive logic: inductive logic.

Deductive logic tells us when the conclusion definitely follows from the premises. That after all is what validity, deductive validity, captures but when we are trying to figure out what the world is like, we are rarely if ever in a position to deduce new theories from solid evidence. Sometimes, we deduce them from speculative theories and find ourselves in a position to test those speculative theories because of that deduction. When we develop our theories, we normally commit to them, either for the purposes of further research or even more strongly than that, well before all the evidence is in. In fact, it is not clear what "all the evidence is in" could really mean. After all, any decent scientific theory will always apply to new cases. That is what makes the theory useful. The theory is not just summarizing the evidence we have; it goes beyond that evidence to new cases. What is needed is an account of less than absolute support. Inductive logic fully formed would be a theory that would measure the degree of support any given premises give to a conclusion. We roll the die five times and a 6 does not come up. That gives some evidence to the theory that the die is unfair. However, if we roll the die five hundred times and a 6 does not come up, that intuitively gives more support to that theory.

There is no question that we think we understand the notion of evidence that confirms a general theory. The theory that all ravens are black, for example, is supported by the observations made of black ravens. The more black ravens we see, the better confirmed the theory is. Seeing white shoes or red roses, does not.

There are then two principles that seem intuitively attractive about confirmation. The first is that logically equivalent theories are confirmed by the same evidence. After all, if a theory R is logically equivalent to another theory S then they are true and false in exactly the same situations. Thus, anything that confirms and is evidence that one is true should be evidence that the other is true. This is often called the equivalence condition. The theory that all ravens are black and the theory that all nonblack things are not ravens can be expressed formally as that $\forall x(Rx \rightarrow Bx)$ and $\forall x(\neg Bx \rightarrow \neg Rx)$. In addition, these are logically equivalent theories.

Another intuitive principle, witnessed in the discussion of the way observing a black raven confirms the theory that all ravens are black, is that for any predicates, B and R and name c, the truth of Rc and Bc confirms the theory that (all Rs are B). More formally, $\forall x(Rx \rightarrow Bx)$.

This condition is often called the Nicod Condition.

It is a matter of classic logic that $\forall x(Rx \rightarrow Bx)$ is logically equivalent to $\forall x(\neg Bx \rightarrow \neg Rx)$. Therefore, by the equivalence condition, whatever confirms one confirms the other too. However, a white shoe confirms the second, so it must confirm the first too. Observing a white shoe confirms the theory that all ravens are black. This is surprising to say the least.

Do we understand confirmation? There are two responses I will consider below. One is a rejection of the idea of confirmation as a key theoretical concept, and the suggestion that the experiment should not be geared toward getting confirmation of our theories. Rather, we should be designing our experiments as attempts to test, or refute, our theories. This is Karl Popper's move. The other response is within the general confirmationist perspective but which suggests alternative ways of cashing out the notion of confirmation. The most common of these is the Bayesian approach—an approach to updating confidence in our theories.

C. G. Hempel (1945a,b), however, noticed that the intuitive ideas about evidence above led to trouble. The theory that

All ravens are black.

is logically equivalent to the theory that

All nonblack things are nonravens.

This is just a slightly more verbose way of expressing what the original theory expressed. We have seen that observing black ravens is intuitively confirmation of the first theory. What is intuitively evidence for the second theory? Well, because it says that all nonblack things are nonravens, it would be a nonblack thing which is not a raven. Thus, for example, a white flower would be evidence that all nonblack things are nonravens. However, if a white flower is evidence that all nonblack things are nonravens and that theory is logically equivalent to the theory that all ravens are black, then the white flower is evidence that all ravens are black! This is bizarre and very unintuitive. There seem to be two types of responses. First, we could accept that white flowers are, albeit surprisingly, evidence that all ravens are black. Or we could deny that the two theories are logically equivalent. Or we could deny that logically equivalent theories are confirmed by the same evidence. Or we could deny that we really have a workable notion of confirmation. The first response does seem to have been Hempel's.

The intuitive idea is that the observation of a black raven is evidence that

All ravens are black.

Observing more and more black ravens is, intuitively, gathering more and more evidence for the theory. On the other hand, were we to observe red pencils, blue pens, white flowers, or purple rain, we would not have discovered evidence for that theory. A quite plausible idea is that if two theories are logically equivalent, then anything that is evidence for one is evidence for the other. After all, logically equivalent theories are true in exactly the same circumstances: they say the world is exactly the same way.

Quine, in his consideration of the Hempel paradox, suggested that the fault lies in the too generous nature of the Nicod Condition. As stated, it was phrased simply in terms of predicates. Quine suspected that the theories relevant to the issue of confirmation did not use just any predicates but specifically depended on the use of natural kind terms. In that case, whereas being a raven may well be a natural kind, being a nonraven or being nonblack is not likely. If the Nicod Condition is restricted so that it only applies to predicates that express natural kinds, then the white shoe will not confirm the theory that all nonblack things are nonravens and, for that reason, the white shoe will not confirm the theory logically equivalent to the theory that all nonblack things are nonravens.

In fact, the Nicod Condition has other quite surprising consequences. As originally stated, it applied to any predicates, B and R and name c, and the truth of Rc and Bc confirms the theory that (all Rs are B). However, there is a symmetry to the condition so that Rc and Bc also confirm (all Bs are R) simply by reversing the order of the predicates. Therefore, a black raven will confirm that all black things are ravens just as it confirms that all ravens are black. This surprising consequence applies no less to Quine's restricted Nicod Condition than to the original unrestricted Nicod Condition.

THE BAYESIAN RESPONSE

Another response to the paradox of the ravens is to go probabilistic. What should your confidence be in a theory given some evidence? Well, one view is that it should be governed by a rule deduced by an early thinker, Thomas Bayes, when what he wrote about probability was published posthumously in 1764:

Bayes's rule: $P(A/B) = P(B/A)P(A)/P(B)$

This is quite general and thus gives a result for any A and B, as long as $P(B)$ is not zero. This rule is indeed derivable from principles that govern probability. However, it is not obvious that the significance given to the rule is indeed correct. For example, being consistent with the rule is required if we are going to be rational in the sense of being consistent with the probability calculus and not vulnerable to a priori undermining. However, it is not clear that the only way to do that is to update our commitments by conditionalizing.

In this case, the probability that all ravens are black on the evidence that Sparky is a black raven is exactly the same as the probability that all nonblack things are nonravens on that evidence. Because, as we saw, those two theories are logically equivalent. However, this means that a Bayesian approach is not going to distinguish between the way Sparky the black raven is supposed to be evidence for one but not the other theory.

This means that both the Bayesian approach and the inductive logic approach fail to actually show us a way to spell out an intuitively acceptable account of when an observation is evidence for a theory. That may be because the intuitive account of confirmation is actually misguided. There may be various ways to save many of our intuitions about confirmation without saving all of them consistent with a Bayesian approach. In fact, one way to understand modern Bayesian approaches to evidence is just that: to see them as replacing the intuitive and incoherent naïve conception of evidence with replacement concepts none of which has an especially strong claim as being the best account we must have had in mind.

KARL POPPER'S DEMARCATION OF SCIENCE

Popper argued that this shows that confirmation is in a bad way. Some suspect he is right but others have not been universally convinced. Certainly, the attempt to develop an account of confirmation that was as rationally compelling as first-order logic has failed. The idea that logically equivalent theories ought to be supported by the same evidence seems incontrovertible but not much else does.

Popper responds to the problem of confirmation by suggesting the error is in thinking that confirmation makes good sense and that confirmation has no place in an understanding of science. The various attempts to converge on the unique sharpening of the intuitive story about confirmation misleads because it presents science as though it were an attempt to adopt theories that are confirmed by evidence. Popper developed a radically different approach to the nature of scientific inquiry. Science is not about trying to confirm your theory but rather trying honestly to refute your theory. A theory that survives honest attempts at refutation, attempts that might have yielded falsifying evidence but did not, is a good theory. A theory that could not have been refuted might play a role as a metaphysical background for scientific theorizing but cannot be thought of as a scientific theory. Science then is a series of bold conjectures and clever refutations. Science is, from this perspective, constantly in a state of possible refutation. That a theory has survived the attempts to refute it thus far is no reason to think that it will survive the next attempt. Theories do not get more and more confirmed. There may be a psychological tendency to gain confidence in a theory that has survived vigorous refutation attempts but it has no rational foundation. The theory is not becoming more probable. This claim of Popper's cannot use a subjective account of probability, for that is just a measure of confidence. For that reason, quite often, it seems the Bayesians and Popperians were talking past each other.

For Popper, there is subjective or psychological confidence in, for example, the claim that the sun will rise tomorrow. No matter how many sunrises we observe, to believe that tomorrow the sun will rise is a projection beyond the evidence. David Hume had argued that there is no logical justification to believe that claim. For Hume, such a belief spoke to our natural tendencies rather than our cogitative powers. In this, Popper was influenced by Hume. He says "I approached the problem of induction through Hume. Hume, I felt, was perfectly right in pointing out that induction cannot be logically justified." (Popper, 1963, p. 55) Popper's interesting twist is to deny that anything like inductive justification was necessary for the rationality of science. Science is an ongoing sequence of theory development and testing and does not require inductive justification.

From this perspective, the virtues we seek in a theory derive from this seeking of refutations. Among these virtues are that the theory be as simple as possible and also falsifiable. For a theory to be falsifiable, we need to be able to specify ahead of time what would count as falsifying evidence. Therefore, why is the theory that all ravens are black falsifiable? It is so because we can specify what would refute it, namely, a raven that was not black. Notice that the very same object would refute the theory that all nonblack things are nonravens. This makes sense of the fact that these really are the same theory in different guise.

Some Objections to Popper

1. The observations are theory-laden and so cannot be used to falsify the theory in question.
2. Accepting that an observation has refuted a theory is just as "inductive" as accepting that an observation has supported a theory.
3. There seem to be some scientific theories of a universal character that Popper's falsification criterion would seem to rule out.

The Theory-Laden Nature of Observation

This has always struck me as an odd objection to Popper's theory of science. Historically, it is odd because Popper was one of the first to emphasize the idea of observations not being theory-free in his early statement of his account of science, *Der Logik der Forschuung*, in 1958. Far from ignoring this feature of observation, Popper was the person who emphasized it. Therefore, the objection is a little bit disingenuous.

On the other hand, the objection from the theory-laden character of observation would be concerning if "theory-laden" had always meant "theory-laden with the theory being tested." However, of course, it does not. An example suffices to show what is going on. In 1919, Arthur Eddington and his colleagues set out to test Einstein's then recent theory of relativity. Newton's theory predicted that light going past a massive body would be slightly deflected. In 1915, Einstein had argued that the Newtonian deflection would be half of the degree of deflection to be observed according to the theory of relativity. However, how does one observe this? At the time he derived this prediction, he lamented that there was no planet nearby much more massive than Jupiter that would allow for the test to take place. Then, it occurred to him. The answer was to wait for a solar eclipse. The position of the stars is well known and predictable from the night sky. During an eclipse, both Newton and Einstein predict that the apparent position of stars close to the sun in the sky will be changed, and each theory predicts it will change by a different amount. To make these observations, Eddington travelled to Brazil and had to use telescopes. However, is this theory-free observation? Clearly not. Telescopes were built in accordance with theories of the refraction of light through different media. However, the theory that telescopes were built in accordance with is much more general than either Newton's or Einstein's theory. Therefore, although the observations were not theory-free, they were not presupposing the particular theory they were being used to test. Incidentally, it was Einstein's predictions that exactly fitted Eddington's observations. This surprising result was the first big result that made Einstein's theory not just beautiful but plausible.

THE "INDUCTIVE" CHARACTER OF FALSIFICATION

What about the observation that scientists do not simply reject a theory when it has a failed prediction? Doesn't that show that Popper is wrong that falsification is any different from induction? Well, no. Theories do not encounter the world on their own to be rejected or not by themselves. Rather it is, as W.V. Quine emphasized, collections of theories, auxiliary hypotheses, and measuring devices that encounter the world together as a corporate entity. When our predictions fail, we know that either some theory is false, or some auxiliary hypothesis is false, or some measuring device has not worked correctly. In addition, as Quine correctly emphasized, we are free to make accommodations anywhere to try to fit the observation into the corporate entity. When Uranus's orbit did not accord with Newtonian theoretical predictions, scientist looked around for the explanation. They did not reject Newtonian dynamics on the basis of the false prediction. Rather, they rejected one of the auxiliary hypotheses: the hypothesis that there was no other body interfering with Uranus's orbit. They calculated that the observed orbit would be predicted if there were a body of such and such a size in such and such a location. They looked and found Pluto, now not regarded as a planet but still important for its influence on the planet Uranus. In this case, it was not Newtonian dynamics that was regarded as false but the auxiliary hypothesis. In other cases, there have been problems with measuring devices.

The point is this: Popper seems to have the logic of science right. When predictions go wrong, something has to give. If a theory, together with some other stuff, makes a prediction and that prediction turns out to be false then either the theory or some of the other stuff is wrong or we made a mistake in observing whether the prediction came out. Something has to be rejected. There is no rule book that tells us what to reject. Science is not a rule-governed activity like that. It is an activity of trying desperately hard to figure out what to reject when the world does not fit our predictions: as astronomers were to discover. Buoyed by their success in predicting the existence of Pluto, astronomers noted that Mercury, the planet closest to the Sun, was not following the orbit predicted by Newtonian dynamics. Some astronomers predicted that there should be another body, a body they named "Vulcan." They suggested that Vulcan's gravitational effects explained the deviations of Mercury's orbit. In doing this, they followed just the same strategy that had worked with Uranus and led to the discovery of Pluto. Here, however, it failed. There was no such body causing the deviations in Mercury's orbit.

There was nothing irrational in making the prediction that there would be a body causing the deviations in Mercury's orbit rather than rejecting Newtonian dynamics. It simply turned out to be just wrong. Science is not about being right, it is about being wrong in ways we can figure out are wrong. That is, just as Popper emphasized the demarcation of scientific theories from nonscientific is not that the scientific theory is true, it is not that only the scientific theory is meaningful, it is that we can conceive of an observation that would lead us to rejecting it, that the scientific theory is falsifiable.

THE LOGICAL PROBLEMS WITH FALSIFICATION

Hempel argued against Popper by producing examples of universal theories that Popper's criterion of falsification would seem to rule out as being scientific theories. One of the clearest examples Hempel provided was this:

> For every metal there is some temperature that it melts at.

This seems a good scientific theory. It is also universal in character so Popper's criterion should apply and yet there is no observation that would falsify it. To falsify it, you need a particular metal and then the observation that it does not melt at any temperature. However, no observation can give you that information. Suppose we have a metal and, at all the observed temperatures, the metal does not melt. We cannot have examined all the temperatures, there are infinitely many of those, so we have not observed that the metal does not melt at any temperature. Hempel's example uses the fact that the negation of the theory gives us a universal negative claim. He is right that Popper's account seems to go wrong in these cases. After we examine the way Popper and others applied the criterion of falsifiability to Darwin's theory, we will be in a position to see a very plausible response to Hempel's logical problem with falsificationism.

This criterion of Popper's has a lot going for it and some of the most common objections seem misguided but I want to show that although well-motivated, Popper's approach is not quite right. What it gets right is that, in science, we try to make ourselves vulnerable to the evidence. Additionally, thinking of theories as something we are trying to find out are false changes the way we approach the world. That much seems right. Bold conjectures and honest attempts to refute are, in general, a mark of rationality, and so understood, science is just a condensation of our ordinary grasp on rationality. To that regard, I take Popper to have made very valuable contributions to our understanding of science and rationality in general. Falsifiability is indeed something to seek. It is not however the *sine qua non* of science.

IS EVOLUTIONARY THEORY FALSIFIABLE?

The question we started with "Is evolutionary theory science?" takes on a Popperian guise as "Is evolutionary theory falsifiable?" but this is too coarse grained to be answered. There are many different evolutionary theories we can consider and we can ask of each of them whether it is falsifiable.

- Is the theory that evolution has taken place falsifiable?
- Is the theory that the human chin evolved falsifiable?
- Is the theory that evolution has involved natural selection falsifiable?
- Is the theory that the human chin evolved through natural selection falsifiable?

The theory that evolution has taken place is a historical theory. It says that the biological world has changed over time. Is it falsifiable? To be falsifiable is to be able to specify an observation that would refute the theory. It might be thought that this would not be falsifiable. However, a moment's reflection should convince you that you can imagine possible evidence that would refute the theory that evolution has taken place. In this case, it is easy to find such a piece of possible evidence: what if the fossil record never revealed any changes? Were the fossil record to be just like the present organisms, then the theory that there has been biological evolution would be falsified. In fact, the fossil record does show that at different times, the biological world has been exemplified by different organisms. Therefore, this theory is in fact falsifiable. We can specify refuting observations that would have led to the theory being rejected. The observations did not take place but that does not mean that the theory is not falsifiable but only that it is not (yet) falsified.

What about the theory that the human chin evolved is falsifiable? This theory is also a historical theory. It says that at one time, organisms existed with no chins and that later descendants had chins. There are issues with deciding which types of organisms are ancestral to which, but one thing we could predict to have seen in the fossil record that would lead to this theory being rejected is if all species in the lineage allied with the hominid lineage had chins. Indeed, we could have falsifying evidence were chins to be found in the hominid lineage regardless of whether they are directly ancestral to our species. Just as we might say that there have been changes to the heart in humans were we to find that humans evolved features or structures novel to their hearts; we do not say that the heart evolved in humans when all vertebrates have hearts. So the theory that the chin evolved in humans is falsifiable. Again, there are observations we could have observed that would have told against that theory.

It is important to note that saying that a theory is falsifiable does not mean that if it were false then there would in fact be some observable evidence that tells you the theory is false. There may be many circumstances in which you cannot tell whether a particular theory is false and only a few in which you can tell the theory is false. Such a theory is nevertheless still falsifiable.

What about the theory that natural selection has been involved in evolution? This is a historical theory, once again it is about what happened in the past but it is not just about what happened, it is about why what happened happened. Saying that something evolved by natural selection involves identifying the mechanism that drove that change. Is that visible? It is important to see that it is not visible. In fact, there is a general issue with causation.

What we aspire to in science is more than a just a catalogue of what happened when; although finding out what happened when is often hard enough. In scientific endeavors, we do try to discern why things happened: the causal explanations. Since Hume, we have known that this is very difficult. We see constant conjunctions easily enough but finding causation is much more difficult, or rather, getting clear what we need to establish to have established the causal link is the difficult question. One idea, coming from Hume himself, is about what happens when we manipulate the system. Hume famously gives two versions of the nature of causation. The first is entirely empirical: it is that A causes B just when A-type events are constantly conjoined with B-type events and A and B are spatiotemporally contiguous and we come to expect B when we perceive A. The second account involves counterfactuals and suggests that A causes B if they are spatiotemporally contiguous and manipulating A manipulates B. Therefore, were A not to occur, then B would not either.

Exploring the science, metaphysics, and epistemology of causation is not a small topic. It may be enough to say that finding the constant conjunctions is hard enough in science, finding causal relations is harder still.

However, here a strange fact needs to be highlighted. This theory that we have decided is not falsifiable is in fact that conclusion of Darwin's key argument for natural selection. Darwin argued for this conclusion from two premises that we saw were empirical and falsifiable. The first, that there is an excess of young relative to breeding individuals. The second, that there is variation and that the variations are heritable. Each of these was empirically testable even in Darwin's day and his evidence for both was copious. Darwin also used another premise that amounted to an analytic claim: If any variations help their bearers in the struggle for existence, then these will be fitter and have a higher tendency to survive. From these three, together with a law of large numbers-type of hidden premise, he derived the conclusion that natural selection would have taken place.

This argument is just like the argument we could have given concerning the die with one red and five blue faces and no other difference we can discern in the die. We could reasonably expect that if this die were to be rolled once a minute for a hundred years that more blue faces than red faces would turn up and that because of the features we advert to: the number of faces colored red and blue. It is reasonable to think so. It is not impossible that we spend the hundred years doing this and find that we were wrong to think so. It may be that in that hundred years of rolling of that die only red faces have turned up. That is not impossible but it is unlikely.

LESSONS ABOUT FALSIFICATION AND SCIENCE

Popper changed the way scientists looked at their theories and their understanding of experimental design. In that regard, he may have been the most influential philosopher writing on science of the twentieth century. Additionally, his influence seems to have been longer-lasting on the scientists themselves than on philosophers. That influence on scientists was, on the whole, salutary. The idea that a good scientific theory ought to be the sort of thing that we could discover to be false, were it false, is the sort of simple idea that changes actual scientific practice. This was a guiding idea, for example, in the debates about the development of biological systematics in the period from about 1965 to 1985. Similarly, the idea that experiments should not be thought of as attempts to confirm our theories to be correct but rather as honest attempts to refute our theories, changes the way we design experiments. Having read Popper, we no longer think of experiments as an attempt to show our theory to be true but as an opportunity to find evidence against our theories by looking to predictions made by those theories. However, there were limitations in his approach, which led to misguided mudslinging among scientists. The idea that an experiment or observation is an attempt to refute a theory and so had better be conceivably going to do that is a good idea. Otherwise, the experiment is waste of time. If nothing you could conceivably measure could tell against your theory in the experiment you have designed, then the experiment is not worth doing. However, does that mean that the theory you have is not scientific? Not at all. We have seen that the theory that natural selection is a mechanism of evolutionary change is not falsifiable. After all, this has the form of an existential claim, even if you find that one example were not to be an instance of natural selection, this is no reason to doubt that natural selection is a mechanism of change. What we saw though, is that there is a reason to think that natural selection plays this role and that is Darwin's own argument for natural selection. That argument suggests that *if* some heritable variations are helpful, *then* such variations will on the whole come to predominate in the population. Notice that this does not single out which variations are helpful, nor does it explain why such variations are heritable. It merely says were such heritable variations to arise, they would tend to be passed on because they are helpful so disproportionately more of their bearers would survive to reproduce and they are heritable so they would tend to be passed on.

The lesson of Darwin's theory for the Popperian approach to science seems to be this: Popper errs by asking whether a theory is falsifiable (and therefore part of science), only considering it as a stand-alone sentence. A theory can be unfalsifiable although it follows logically from empirical (that is, falsifiable) claims and analytic claims. Because it is derivable from such claims, it is certainly a commitment anyone holding to those premises must accept on pain of contradiction. When the theory has the explanatory power of Darwin's theory, it becomes clear that Popper's mistake is to ignore the justificatory structures that theories can come with. An unfalsifiable theory can sometimes be deduced from unimpeachable premises. An extended sense of the scientific is required. Whether we do that by extending our notion of falsifiability or just accept that the scientific encompasses the narrowly falsifiable and what follows appropriately from those and analytic truth is of no real consequence. The effect is the same: the theory of natural selection is a good scientific theory. Extending the notion of falsifiability to encompass structured theories as presented in Darwin's key argument would allow us to see the scientific status and virtues of Darwin's theory of natural selection.

What now of the theory that the human chin evolved by natural selection? This is, in many ways, the pointy end of the theory: its application to the world.

There is a tendency to argue like this and I am caricaturing here but biologists will recognize the moves.

Feature F is present in these organisms
So it evolved
So it must have evolved by natural selection
So why did having F make those organisms fitter than those without F?

The observation of F in those organisms is perfectly empirical and justified. The claim that it evolved is, given the reality of biological evolution, also justified. However, thereafter, we are in the realm of leaps in the dark. Even Darwin did not think that everything that evolved did so by natural selection. In that case, the inference from F evolved to F evolved by natural selection needs motivation. Even supposing that F evolved by natural selection, having F might not be why the organisms with F are fitter. Organisms are not modular to that extent. We cannot simply add F, typically, and leave other features unchanged. Even if we had a guaranteed way of knowing which evolutionary changes were due to natural selection, knowing that would not guarantee that we knew why those that were selected were selected. That is, how was it that these were fitter? The variety that was fitter differed in numerous features from the less fit variety. Which features actually made the difference in fitness and which features made no or negative differences in fitness?

When it comes to the chin and its evolution, we have numerous attempts to give explanations for the novel feature *Homo sapiens* has and other close relatives such as *Homo neanderthalis* and *Homo heidelbergensis* lack. When we do that, we are proposing particular abilities that are conferred by the chin. Perhaps the receding tooth-bearing dentary zone relative to the bony mandibular zone makes for more efficient grinding? Or perhaps ... Each of these explanations is an attempt to find a way that the feature in question has a causal power that differs from the ancestral condition. Moreover, it is not enough to find a difference in power, there is also a historical claim that this difference is what explains the success of this variety over the alternatives. However, where there is no difference in causal power, or where the supposed advantages do not accrue from the difference, the explanation of the historical change is stymied. Holton et al. (2015) intriguingly have argued that the proposed mechanical advantages to the chin do not actually bear out when we look at the physical structure and forces involved in using the jaw. Mechanical forces could not explain the origin of the chin. In fact "Individuals with the most mechanical resistance had chins most similar to a 3- or 4-year-old—meaning they didn't have much of a chin at all." Supposing Holton and his colleagues are correct in this, they will have shown that one selective explanation for the origin of the chin, the advantage accrued mechanically in dealing with certain forces, is wrong. This does not show that other similar accounts are false. However, this does show why it is that we are never in a position to say that the claim that the chin arose through natural selection for the chin is false. Such claims look to be impossible to falsify. This is the serious complaint that Gould and Lewontin (1979) raise to adaptationist approaches to evolutionary explanation. We can never specify what it would be to be in a position to say that the chin is not an adaptation. This paper is one of the most read papers in evolutionary biology and one of the most criticized. However, it has lessons to teach and should be read closely and carefully. The positive account in that paper was controversial and speculative and served as a distraction from the main points that would have stood out more clearly without that speculation. Gould was very interested in what he called "structural constraints." Actually, whether there are such things and whether they are a real limitation to natural selection seems to be largely irrelevant to the question of optimality. Were there constraints then they would be reflected in the shape of the fitness landscape: what might have seemed to have been an available phenotype turns out not to be. The question would remain whether the population we are examining is at a global maximum, a local maximum, or neither. The issue of optimality is not settled either way by the existence or the nonexistence of structural constraints.

Therefore, the problem that remains seems to be this: adaptationism is something we have good biological reason to believe is false. Evolutionary biologists nevertheless start thinking in terms of interpreting features as adaptations, and that as we have seen is a never falsifiable claim. However, what is not emphasized often enough is that even if the feature is an adaptation, it is another inference, unjustified without further grounds, to the conclusion that the feature has its origin due to selection for that adaptive significance. Whether a feature is adaptive is a question of the consequences of having that feature for the variety. The claim that the features' presence is due to selection for that feature is a causal historical claim. Certainly, if we knew that it was adaptive, we would know that it is available to be selected but that is not guaranteed. Moreover, as we have seen already, being adaptive requires only satisfying, being good enough, and better overall than the competitors.

The problem Lewontin and Gould leave us with is that we are in no position to know when and how to use adaptive explanations and when to use alternative explanations. As we saw, part of what Darwin wanted to explain is the fact that organisms do fit into their environments. This does not require the theory of natural selection to observe. The problem is when we are trying to tease out the causal history to discover how we got to the point we are interested in, we are not just trying to say what followed what, although that is not easy a lot of the time. We are trying to say why changes took place—but also why things are as they are—and we are in the inevitable situation that we have to conjecture about what changes took place and why. These conjectures, however, seem hard to test, but test them we must if we are to seek that empirical friction that makes science worth pursuing.

SCIENCE AND EVOLUTION: WHAT THE SCIENCE GUY COULD HAVE SAID

Late in writing this book, I watched a video streamed online of a debate* between Bill Nye "The Science Guy," a television personality in the United States and Ken Ham, a transplanted Australian living and working in the United States and the

---

* The debate took place on February 4, 2014. Available at https://www.youtube.com/watch?v =z6kgvhG3AkI.

founder of The Creation Museum. I found this debate interesting but it was not fun. It was enlightening but not encouraging. The topic was focused on what we should say about the age of the universe, the history of the earth and the question of evolution. Nye suggested that we knew quite a bit about these things. We have gone quite a way down the track in making sense of many questions we have been wondering about for a very long time. When challenged about his claims, Ken Ham reiterated again and again "There's this book!" as befits someone who is so closely related to the Creationist "Answers In Genesis" website (https://answersingenesis.org/). Bill Nye, on the other hand, repeatedly claimed that all he was doing was following the evidence and here is where things got interesting.

Ham distinguished between observational science and historical science. He claimed that observational science is something Creationists and Evolutionists can agree about. This is very interesting and is indeed a distinction that needs making. After our discussion of Gosse, however, we are in a position to make some progress in articulating how to understand this debate. First, let us be clear. Ken Ham is right that it is possible to accept all the evidence the evolutionists point to and yet hold to the literal truth of Genesis. The young Earth theory is, after all, just a version of Russell's skeptical hypothesis. So let's consider the situation. We face a choice between accepting that the universe, and the Earth in it, is billions of years old or that there is a special reason to accept an alternative theory. In Russell's case, we saw that we had no reason, it seems, to accept the universe began ten minutes ago rather than eleven minutes rather than fifteen years rather than … This makes accepting Russell's skeptical hypothesis arbitrary.

How does it stand with Ken Ham's basis for the young Earth, namely, the biblical account? Here, things are also unhappily arbitrary. Had we a commitment to the literal truth of the Bible, we could make an accommodation but it is of a desperate sort. Even young Earth creationists accept that stars are many thousands of light years distant. They emphasize that a light year is a distance in space not time. Ham himself made this quite correct observation in the debate with Nye. Using the Hubble telescope, we are able to see an early forming galaxy in the eXtreme Deep Field (XDF). This galaxy (UDFj-39546284) is the most distant body Hubble has discerned at 13.2 billion light years away. Were the universe, as Ham insists, merely 6000 years old, the light coming from such objects could never have been further than 6000 light years from Earth. In that case at Creation, God would have created the world with light already travelling through the Universe. The light that we receive on Earth and that reveals the distant galaxy has not come from that galaxy. This is a version of Gosse's *as if* history.

Of course, ad hoc assumptions can come in to save the literal truth of the Genesis account of creation but they cannot rescue God from the charge of being a deceiver. This is a charge curiously absent from the scientific attack on Creationism but, historically, it was the key charge that led fundamentalists to repudiate Gosse's rapprochement of biblical literalism with the scientific evidence. There are also often appeals to ad hoc assumptions concerning the consistency of the explanatory laws. What if the speed of light behaves differently in different parts of the universe? What if fossils sort seemingly nonrandomly by a law of nature that no longer applies?

The view that knowledge requires certainty was once a driving idea in shaping our understanding of both philosophy and science. Descartes, who ironically is best known today in connection with Cartesian skepticism, argued strenuously that for a belief to count as knowledge, it has to be true and certain. Moreover, far from being a skeptic, Descartes thought that we could know the world and if we followed his argument that took a crucial byway through the proof that God exists and is no deceiver, we can achieve certainty about our scientific beliefs.

The Logical Positivists also sought a rational methodology that given a body of evidence would tell us which theory we should endorse and how much confidence we should have in it. They saw, from their reading of Hume's skepticism, that certainty would not be part of science except for those bits of science that were really about definitions and the logical consequences of the definitions: that is, mathematics and other analytic enterprises, as they saw it. The Positivists thought that verification marked an important feature of science because it was an important feature of language having any meaning as such. In this, they were guided by their conviction that what meaning there is in language is something we bring to it. Language, after all, is a social instrument created and modified by people for certain ends. The Positivists were clear that political ends were often served by ways in which language operates. For them, the critical task of making meaning clear served the political aims of democratizing language. If a bit of language has a meaning, it must be because we have used it in certain ways applying a term correctly, for example, or rejecting the misuse of the term should that occur. This is the way the Positivists understood the idea of meaning is used. Einstein's operationalizing of simultaneity fitted their model perfectly and served as an intellectual beacon. Take a concept that people find intuitive and show that if you spell out the empirical conditions of applying the concept, you find a richer concept: the explication of the concept, in Carnap's terminology.

The Positivists had by the 1970s fallen a long way out of favor. In fact, their reputations had been tarnished to the point that calling someone a Positivist is often taken to be a denigration of that person's views. This is quite misguided and morally bankrupt. The Positivists in fact were the continuation of the efforts of the Enlightenment and their opponents in many cases represented the forces of the Romantic reaction to the Enlightenment. The views the Positivists held maybe false and wrongheaded but, it seems to me, they were well motivated and they made real progress among the mistakes.

Popper's response to the Positivists' problems with confirmation was his criterion of falsification. This did in fact point to a key idea that has motivated science: empirical vulnerability. Popper claimed that what is distinctive about science is that it seeks the friction with the world that comes with empirical vulnerability. He treated that too simply by considering sentences in isolation and asking whether as isolated sentences they were falsifiable.

The big lesson to draw from this chapter is that science is not like that. Quine (1953) said that it is the whole of science that encounters experience. We do not need to go that far. We saw Darwin's key argument in Chapter 4. In addition, what is interesting about this argument is that Darwin derives his theory of natural selection from some empirical claims (and they are measureable and so falsifiable on their own) and some analytic or quasi-analytic claims (the definition of fitness and the law of large numbers). Therefore, although the statement of his theory is not falsifiable, it follows impeccably from falsifiable and analytic truths. The idea that such a theory should be other than within the province of science reflects badly on the criterion of science and how it is deployed. Darwin's theory is science. In fact, it is one of the best exemplars of science.

REFERENCES

Curd, P. 2007. *Anaxagoras of Clazomenae: Fragments and Testimonia.* Toronto: University of Toronto Press.

Edelstein, L., ed. 1943. Hippocrates. In *The Hippocratic Oath: Text, Translation and Interpretation.* Baltimore, MD: Johns Hopkins Press.

Goldacre, B. 2012. *Bad Pharma.* London: Faber and Faber.

Gosse, P. H. 1857. *Omphalos: An Attempt to Untie the Geological Knot.* London: John Van Voorst.

Gosse, E. 1907. *Father and Son.* London: W. Heinemann.

Gould, S. J. and R. C. Lewontin. 1979. The Spandrels of San Marco and the Panglossian paradigm: A critique of the adaptationist programme. *Proceedings of the Royal Society of London, Series B* 205: 581–598.

Hempel, C. G. 1945a. Studies in the logic of confirmation I. *Mind* 54(213): 1–26.

Hempel, C. G. 1945b. Studies in the logic of confirmation II. *Mind* 54(214): 97–121.

Holton, N. E., L. L. Bonner, J. E. Scott, S. D. Marshall, R. G. Franciscus, and T. E. Southard. 2015. The ontogeny of the chin: An analysis of allometric and biomechanical scaling. *Journal of Anatomy.* DOI:10.1111/joa.12307.

Popper, K. 1963. *Conjectures and Refutations: The Growth of Scientific Knowledge.* Routledge and Kegan Paul.

Popper, K. 1976. *Unended Quest: An Intellectual Autobiography.* Glasgow: Fontana/Collins.

Popper, K. 1978. Natural selection and the emergence of mind. *Dialectica* 32: 339–355.

Popper, K. 1985. Natural selection and its scientific status. In *Popper Selections,* edited by D. Miller. Princeton: Princeton University Press.

Quine, W. V. O. 1953. *From a Logical Point of View.* New York: Harper Torchbooks.

Ruse, M. 2000. *Can a Darwinian be a Christian?* Cambridge: Cambridge University Press.

Ruse, M. 2006. *Darwinism and Its Discontents.* Cambridge: Cambridge University Press.

Russell, B. 1922. *The Analysis of Mind.* London: George Allen & Unwin.

Tsantsanoglou, K., G. M. Parássoglou, and T. Kouremenos. 2006. *The Derveni Papyrus,* edited by S. Leo. Florence: Olschki Editore.

7 Heritability of Characteristics
The Role of Genetics and Epigenetics

Darwin's theory is founded on a number of key notions: fitness, variation, and heritability among them. The heritability of a characteristic is crucial if that characteristic is to play a role in the process of evolution by natural selection. Just as the product, evolution, can easily be confused with one of the mechanisms that produce evolution, natural selection, it seems the heritability of a characteristic is easily confused with its being one of the mechanisms that underpin heritability and genes. In this chapter, we will see why this is an important mistake. What we will see is that there is a very important sense in which English, or the mastery of it, is a biologically heritable trait. This claim sounds wrong and has commonly been controverted. In the literature on the evolution of language, it is a commonplace observation that unlike the language faculty itself, speaking this or that language is not heritable. Showing that English is, strictly speaking, a heritable characteristic is useful because it highlights the mistaken assumptions that might lead some to think it is false.

The recognition of our biological nature is becoming more and more widespread and pervasive. Even areas such as literary criticism, linguistics, ethics, and reasoning are not immune from being reappraised as bearing deep and important imprints of that biological history. In this context, however, there is a tendency to think we can distinguish between a heritable genetic component in our cluster of abilities and an environmentally determined nonheritable component. In this way, we find a common contrast between the speaking of a particular language, say English, which is taken to be environmentally determined, and the possession of the cognitive architecture, which facilitates that language acquisition and is taken to be a genetic inheritance. The interest in the example of the mastery of English is not so much in the heritability of English as such but in the mistakes often associated with the strict biological notion of heritability and, importantly for us, how this can be misleading about the very process of evolution by natural selection. It is commonly held that language universals, for example, might be heritable and so genetic, whereas speaking this or that language as a first language is certainly not genetic and so not heritable (Anderson and Lightfoot [1999]; Hurford [1999]; Dunbar [2000]; Knight et al. [2000]). That argument, and the way of thinking it exemplifies, contains a number of illuminating mistakes. The way I shall show this is by showing first that the mastery of English is, strictly speaking, a heritable trait and then argue that those who dispute the claim that the mastery of English as a first language is heritable do so for bad reasons and often thereby betray a misunderstanding of the strictly biological notion of heredity. The most common reason for disputing the thesis is an illicit equation of heritability and genetics, as exemplified in the preceding argument. So let us be clear: the thesis is not that the mastery of English is a genetic trait, nor is it that it is "in the genes," whatever that vague and misleading phrase may mean. In saying that much, I should not be understood as arguing for any deviant biological theory, such as what passes for so-called neo-Lamarckism. There may be cases of somatic changes that induce genetic changes, or not, and my thesis is unaffected. It is probable that any biologist who hears the arguments here would think that nothing radical has been said. The thesis that English is heritable and in general is inherited certainly does not depend on any deviant biological theory. I use this clear example because the common reaction to the thesis that the mastery of English is heritable is a good example of the way misunderstandings of evolutionary theory and biology in general have led otherwise sensible researchers into the mire. To understand what is involved in the claim that the mastery of English is a heritable trait, it is important to go over the key notions involved and once that has been done, the major thesis of this paper will simply fall out of the discussion.

GENOTYPES AND PHENOTYPES

Let us then begin with the distinction crucial in this area: that between an individual's genotype and the individual's phenotype. Roughly speaking, the genotype is the genetic endowment of the individual very narrowly construed, and the phenotype includes all other traits of the individual. Dawkins famously suggested extending the phenotype to any characteristic caused by the genotype. We will examine that suggestion below. Together, the genotype and phenotype of an individual constitute all that individual's traits. Therefore, why do we make the distinction between genotype and phenotype? It will transpire that there is a crucial relationship between an individual's genotype and phenotype.

The genotype is that individual's genetic inheritance. Exactly what the genetic inheritance amounts to is an interesting biological question. Is it simply which alleles the individual has? Is it the arrangement of the nucleotides on chromosomes? Should it include mitochondrial DNA as well? Is the extent and character of the DNA methylation part of the genotype? Not much depends on the particular details of the answers to these questions and, in these answers, we should be guided by the theoretical needs of biologists. The notion of phenotype will march in step with whatever notion of genotype we fix on. The crucial thing is that the genotype and the phenotype are correlated notions. Genotype and phenotype together encompass all of an individual's characteristics.

The phenotype includes such characteristics as having a head, having five fingers, being a male, speaking English, and voting in the 2015 election, but also traits at the level of chemical structure: for example, just which amino acids an individual's cells contain, and in what concentrations, which proteins they synthesize, and with what tertiary structure. Phenotypic traits need not be consistent over time. Speaking Italian, or weighing less than 100 kilograms, is a phenotypic trait an individual may have as a four-year-old and lack as a forty-year-old.

The notion of the phenotype is an inclusive notion. It includes all traits of an individual that are not in the narrowly conceived genotype. This means that if that individual synthesizes a particular enzyme, then that is a phenotypic trait of that individual. It means that if that individual has one head and four limbs, then that is part of the phenotype of that individual. In some extended sense, the property of being in a world in which a president with the initials "JFK" was assassinated is a property of all individuals in the world. However, we do tend to limit our understanding of the traits of the individual to those which are genuinely properties of the individual more narrowly conceived. One such phenotypic trait narrowly conceived is the speaking and understanding of English. This is, on all counts, a phenotypic trait of an individual.

THE FUNCTIONAL RELATIONSHIP BETWEEN GENOTYPE AND PHENOTYPE

The interest in making a distinction between the genotype and the phenotype is in the relationship between them. The important idea to grasp here is that the phenotype is the expression of the genotype relative to the environment. In other words, the phenotype is functionally dependent on the pair of a particular genotype and a particular environment. Given a particular pair of a genotype and an environment, that particular phenotype is expressed.

$$\langle \text{Genotype 1, Environment 1} \rangle \rightarrow \text{Phenotype 1}$$

A phenotype must be conceived as the outcome of the coming together of a particular genotype and a particular environment. This means that the relationship of genotype and environment to phenotype is functional in the mathematical sense. A genotype–environment pair maps onto a phenotype. That same phenotype may be the value of various other genotype–environment pairs as well. In addition, for each genotype–environment pair, there is a phenotype onto which that pair maps.

Notice that this relationship should not be expressed as an equivalence. It is not that

$$\text{Genotype 1} + \text{Environment 1} = \text{Phenotype 1}$$

That claim is nonsensical; the problem with this is that it asserts that the particular phenotype is the sum of a particular genotype and environment. However, that makes no sense. A phenotype is not the same thing as the sum (whatever that is) of a particular genotype and environment. The phenotype is the outcome of a process of interaction of the genotype and environment. We can indeed represent the relationship in functional terms as

$$F(\text{Genotype}_1, \text{Environment}_1) = \text{Phenotype}_1$$

where the functor "F" picks out the characteristic function that maps pairs of genotypes and environments to phenotypes. This means that to each pair of a genotype and an environment, a particular phenotype is assigned; intuitively, this is the phenotype that is determined by that pair of genotype and environment. In the same way we might insist that the height of the mercury in the tube is not itself the temperature, we can find a relationship between the height of the mercury and the temperature outside. There is a function, T, which takes us from the height h of the mercury in the tube to the temperature, t. It would be a mistake to say that

$$h = t.$$

After all, one is a height and the other a temperature. It is correct to say

$$T(h) = t.$$

That is, there is function, T, which given a height in the tube delivers a temperature. That is why we can mark the numerals on the tube to indicate the external temperature. Another nice aspect of this example is that we can see that the temperature can depend functionally on the height even though it is clear that the temperature is not caused by the height. The height indicates the temperature but does not cause it. Rather, the temperature (ignoring for now the near consistency of the atmospheric pressure) causes the height of the mercury to expand and contract.

The functional account of the phenotype depends on underlying assumptions about the way the world works at a physical level. In fact, built into the account described thus far is the idea that the underlying physics is deterministic because it assumes that, given a genotype and an environment, only one phenotype is delivered. A more nuanced account would make that relationship a probabilistic one. That is, instead of characterizing the relation between genotype–environment pairs and phenotypes by a deterministic mathematical function, that relationship can be treated as a probabilistic relationship. In the indeterministic case, a particular pair of a genotype and environment makes a phenotype probable to some degree or other. We can afford to ignore this recherché subtlety in what follows. Nothing in the discussion of heredity really depends on that assumption of determinism. All the substantive points would survive *mutatis mutandis* were we to use the more general probabilistic account.

INSULATION: THE CONTRAST BETWEEN "GENETIC" AND "ENVIRONMENTAL" TRAITS

Each individual has a particular phenotype and is the developmental outcome of a particular genotype in a particular environment. However, we are often interested in the relative insulation of particular phenotypic traits (specific aspects of an overall phenotype) to variations either in the genotype or in the environment that in fact brought about that particular phenotype. In that case, we are considering what difference it would make to that phenotypic trait were we to vary the particular genotype or environment. This is explicitly counterfactual but we have evidence in the form of various genotype and environment combinations and the phenotypes they produce. That evidence is, however, quite limited. We never have examples of all genotypes paired with all environments. A phenotypic characteristic is insulated to changes in environment within a range just when (holding the genotype constant) the phenotype is realized regardless of which environment in that range is chosen. Similarly, we can say that a phenotype is insulated to changes in genotype within a range when (holding the environment constant) the phenotype is realized regardless of which genotype in that range is chosen. We can also talk about insulation to changes in both genotype and environment by looking at the various genotype–environment pairs that give rise to that phenotypic characteristic.

The concepts of relative insulation of a phenotypic characteristic to environmental change and insulation to genotypic change are important because they tell us something about the causal mechanisms that lead to that characteristic.

Different phenotypic characteristics or traits are differently insulated to changes in genotype and environment. Understanding how the traits are insulated involves a very comprehensive understanding of the ways in which development of traits occurs. There are some phenotypic traits that are relatively insulated to changes in the environment. The possession of arms and legs would seem to be an example of this sort of trait: over a large range of environments, these characteristics develop (when we hold the genotype constant). It is important to notice that this insulation is not absolute. There may be some variations in the environment, such as the introduction of thalidomide at certain points *in utero*, to which these traits are not insulated. The lesson is that all questions of relative insulation are more or less and that all phenotypic traits require both a genotype and an environment.

Each complete phenotype is in fact the developmental outcome of a particular genotype and a particular environment. An individual actually develops its phenotype in an environment, after all. Every phenotypic characteristic that is part of that complete phenotype is the caused outcome of both. Some characteristics are exhibited over a range of genotypes but covary with environmental features. That is, given an appropriate genotype, whether an individual has the characteristic or not depends on a particular kind of environmental feature. Change that environmental feature and those individuals will not develop that characteristic. For that reason, we could call that characteristic an environmentally determined trait because that trait is not insulated to changes in that aspect of the environment.

Analogously, there are characteristics that, given a range of environmental conditions, depend on the presence of certain genotypes. Given those environmental conditions, whether the characteristic is developed is sensitive to which genotype the individual has; change the genotype and you may not have that characteristic. These are characteristics that are not insulated to changes in genotype. For that reason, we can call those characteristics genetically determined characteristics. However, again this is not an absolute notion. No genetically determined phenotypic characteristic is absolutely insensitive to what the environment is like. Even the forming of protein chains depends on the availability of the amino acids, an environmental feature. The tertiary structure of proteins is determined by the patterns of folding that protein chains undergo. Such folding also turns out to be sensitive to chemical features that are environmental features.

A phenotypic trait that is exhibited across a range of genotypes, but is not insulated to changes in an aspect of the environment, might in short be called an

"environmentally determined trait." Analogously, a phenotypic trait that was relatively insulated to changes in environment but dependent on the possession of a
particular type of genotype might in short be called a "genetic trait." However, these
understandable rough and ready ways of speaking might mislead the inattentive.

Notice that no characteristic will be utterly insulated to changes in environment
or to changes in genotype. The insulation is always an approximate notion. This
just reinforces the observation that development always involves both a genotype
and an environment. However, we can think about what is "normally" the case and
think of characteristics that are relatively environmentally insulated, as opposed to
those that are relatively genetically insulated. *Normally* is a hedge term with no clear
boundaries. What we need to remember is that when we talk about features being
environmentally insulated (and thus genetic) or genetically insulated (and thus environmental), we are speaking loosely when what actually is involved is always the
interaction of genes and environment.

What has just been described as the insulation of phenotypic trait against certain
variations in genotype is a more general form of what has recently been discussed
in the developmental literature as "robustness" and it clearly has connections with
Waddington's notion of canalization. Waddington's notion involves a trait becoming
more insulated to variation in genotype and environment.

Although using this notion of insulation allows us to characterize a sense in which
a trait can be genetic or environmental, it is important to see that these labels are used
in an attenuated sense. A phenotypic trait that might be called "genetic" (because it
is relatively insulated to variation in environment, given the genotype in question)
is nevertheless still the outcome of a genotype and an environment. Moreover, there
may be ranges of environments in which those genotypes do not lead to that phenotypic trait. Similarly, a trait that we can call "environmental" might nevertheless
depend on the possession of one of a certain range of genotypes. In neither case is it
right to say, strictly speaking, that we have shown the lie in the functional relationship displayed in the previous case. It remains true that the phenotype is the expression of the genotype relative to the environment.

In principle, the same phenotype can be the outcome of various combinations of
a particular genotype and a particular environment. Understanding the way the phenotype can be caused is understanding just which sorts of genotypes combined with
which sorts of environments lead to that phenotype.

Traits that are genetic in this attenuated sense by being relatively insulated to
environmental variation are not utterly insulated to environmental variation. There
may indeed be many environments that do not lead to the phenotype, either because
they are fatal and not conducive to development or lead to that phenotype not being
developed. For that reason, it is important to note that this is not the same thing as
saying that the trait is "in the genes."

This leads to a change in perspective. The issue is no longer which phenotypic
traits are in the genes. Rather, given a particular phenotypic trait that developed as a
product of a particular genotype and environment, we are interested in what sorts of
variation in genotype and environment is that trait insulated against.

Sometimes, when the phenotypic trait is undesirable, we are interested in what
sorts of intervention will ensure that a different trait develops. Some interventions

involve changes in environment; some proposed interventions involve changes to the genotype.

When we take this functional approach to the relationship between phenotype, genotype, and environment, we get a function that maps the pairs of particular genotypes and environment to particular phenotypes. This has mathematical consequences that show the assumptions involved in the modeling developed by population geneticists. First, the function is well defined when the relation is a mapping from genotype–environment pairs to phenotypes. That relation can hold even though there is no unique ordering, nor indeed any ordering on the genotype–environment pairs. It would be a substantive claim that there is such an ordering. Moreover, the supposition that there is such an ordering imposes cardinality restrictions we might care to be without. When we think of the multidimensional hyperspace that represents all the dimensions of variation of genotype and environment, we are then mapping that into another hyperspace that represents the different dimensions along which we represent the phenotype. Typically, we do not deal with these hyperspaces at all: they are just too hard to imagine. We collapse each of these multidimensional hyperspaces into one dimension. The ordering imposed by such a collapse might be arbitrary though well defined. Thus, the assumption that the function relating genotype–environment pairs to phenotypes will be continuous is unjustified in general. Notice that here we are talking about properties not of our models of the world but rather of the world. The relationship between genotype and environment pairs to phenotypes is a real-world relation. When we model this real-world relation, we make assumptions for the sake of simplicity; to make the mathematics tractable or to allow us to visualize the relations. We assume that populations are infinite; we assume that we can collapse the multidimensional space that captures the genotype–environment pairs into a single well-ordered dimension, and so on. There is nothing wrong with making these simplifying assumptions as long as we do not make the mistake of confusing features of our model, which are there to make the mathematics tractable, with features of the world.

GENETICS AS A MECHANISM OF HEREDITY

One such mistake is to think that now that we know that genes are a mechanism of heritability, that when we say that a trait, which is a phenotypic trait, is heritable, we are saying that the trait is genetic, even in the attenuated sense I have characterized. To repeat, it is a mistake to think that when we assert that a characteristic is heritable that we are saying that the trait is genetic, even in the sense that we have characterized, namely, it is a trait that is relatively insulated to changes in the environment but not to changes in genotype.

Let us be clear just how attenuated that sense of a trait's being genetic is: As we have characterized it, a trait is genetic in the sense that if an individual had that genotype over a range of environments, then that individual would have the trait, whereas if it had a different genotype, then it may not have that trait. This leaves open just how wide that range of environments has to be and also how different the genotypes have to be. There is no answer to just how wide the range must be. In different contexts, we accept different standards, and it need not be that the range of environments are all possible environments, indeed for most characteristics, there will be environments that do not lead to the individual developing that characteristic; either because those are fatal, or some masking occurs, or.... We typically use notions such as "normal" environments or typical environments to restrict the range of environments.

Analogously, we do not usually depend on functions dominating others across their entire range, that they usually do so is often taken as enough. For example, children with phenylketonuria (PKU) do not produce the enzyme phenylalanine hydroxylase, which metabolizes phenylalanine, an amino acid present in protein (Seashore et al. [1999]; Schweitzer-Krantz and Burgard [2000]; Bailey et al. [2005]). In individuals with PKU, the ingestion of protein that contains phenylalanine leads to the buildup of phenylalanine in the bloodstream and brain tissue resulting in mental retardation. Treatment of this disease often involves a nonprotein diet that arrests the buildup of phenylalanine (Guralnick 1998). This means that the mental retardation characteristic of PKU can be seen as a phenotypic trait that is not insulated to the environmental changes characterizing this treatment regime.

Phenotypic traits differ in degrees of insulation to variation in genotype and environment, not in whether they are in the genes or not. The common turn of phrase is profoundly misleading. Traits are not in the genes the way words are in books nor in the sense that far-off places might be said to be in books in that they talk of far-off places. Apart from capturing the public imagination, the phrase "the language of life" is a sham shorthand for a process that is far more complicated.

HERITABILITY: DIFFERENT NOTIONS

Phenotypic characteristics are the outcome of a causal process that involves the organism's particular genotype but also the particular environment in which that organism develops. Having spelled out that relationship between genotype, environment, and phenotype, we can explore the issue of what it means for a particular phenotypic trait to be heritable. The answer some people offer here, "It is in the genes," is doubly confused.

First, as we have seen, it is confused because the idea that traits are "in the genes" is profoundly misleading about the relationship between genotype, environment, and phenotype. Second, it confuses genes, which are a mechanism of heredity, with the phenomenon of heredity itself. Darwin could readily discern that certain traits were and certain traits were not heritable. As an expert in animal husbandry, he knew full well, as did many, that you are more likely to get a cow that milked heavily from a mother that milked heavily. By drawing attention to the fact that there are variations among individuals and that these variations are heritable, something he could empirically establish, he provided one of the key premises in his argument for natural selection. What he lacked was a mechanism that could explain that heritability. He tried various mechanisms but they tended to lead to the blending of variations rather than their heredity. Therefore, what is the heritability of a trait?

There are three different notions that get picked out by that term (Jacquard, 1983). The first, and core idea, is a matter of resemblance between parent and offspring. This is the core notion in that it is the notion closest to the intuitive notion. For those interested in heritability for the sake of Darwinian explanations, this is the key notion. Spelled out more carefully, the idea is that a character is heritable just when and to the degree to which there is a correlation between a parent having the characteristic and the offspring having it. Another way of putting this is: the probability that the offspring has the characteristic is high given that the parent has the characteristic.

There are two further and distinct notions of heritability defined in population genetics from which that core notion needs to be distinguished. A second notion is a measure of the proportion of phenotypic variation, which is attributable to variation in genotype; this is sometimes called heritability in "the broad sense." A third sense is heritability in "the narrow sense": this is the notion that has been used in animal and plant breeding and that tries to give a measure of the contribution made by each individual gene to a trait. This last notion seeks to measure the degree to which variation in a particular gene contributes to the variation in a particular phenotypic characteristic.

HERITABILITY AS CORRELATION

Suppose we have a character C, which is present in a particular population and the values of the character C are indicated by X. Let π_i be the proportion of the population for which C takes the value x_i. Then, we can easily calculate the mean μ and the variation of the values of C ($\text{var}(X)$) for that population by

$$\mu = \Sigma_i \, \pi_i x_i$$

$$\text{var}(X) = \Sigma_i \, \pi_i (x_i - \mu)^2$$

Suppose further that we plot the value of X for each individual against the values of C for the parent generation. From this information, we can define a degree of correlation between the offspring's value and the value of the father (or the value for the mother or the average of the values of the mother and father). A correlation of 1 indicates that the father's value of C is a perfect predictor of the value of C in the offspring. Therefore, if the value of the father is n so will the offspring's be n. A correlation of 0 indicates that there is no correlation between the father's value and that of the offspring. The information about the father's value gives us no information about the value of the offspring's value. Correlation can obviously be to a degree, the information about the father's value of C is giving us more or less information about the value of C in the offspring depending on the degree of correlation. Notice that, as defined, the correlation is a feature of the population. Generalizing to other populations of the same species, or to the whole species, or even to any bearers (of whichever species) of the characteristic C, is not warranted on the basis of the definition alone. The extent of the correlation in a population that is generalizable across populations, or indeed across species, is an interesting biological question. There is absolutely nothing that tells us what the answer to that question will be in any case or that the same answer would be obtained for different populations and/or species and/or characteristics. Correlation, however, is a perfectly well-defined notion of heritability of a characteristic in a population. There will be some degree of correlation, that is, the line mapping fathers with offspring will have some slope.

Genetic Heritability in the Broad Sense

Suppose, once again, we have a population within which there is a phenotypic character C whose variation in the population is var(X). It is tempting to conceive of that variation as having two sources, the genetic variation in the population var$G(X)$ and variation in the environment var$E(X)$. These might be thought to be related such that

$$\text{var}(X) = \text{var}E(X) + \text{var}G(X)$$

The measure of the importance of the genetic causes of the total variation is what is known as the genetic heritability in the broad sense (h^2). This can be defined by

$$h^2 = \text{var}G(X)/\text{var}(X)$$

Genetic heredity in the broad sense depends on the very strong assumption that there is an additive relationship between total variations of the characteristic C in the population. That is, that there is a way of assigning values to var$E(X)$ and to var$G(X)$ so that the result is var(X). Even if there were, notice that the value of h^2 is defined only for that population, not for that characteristic in general. Therefore, there is no reason to think that when we have a value for h^2 for a population that we have determined a value that applies more generally.

The problem of additivity is this: suppose we now have μ and are considering the effect of each genotype and environment on the value of X, the measure of C. Suppose each genotype g_i could be attributed with an effect γ_i and each environment e_i could be attributed with an effect ε_i. In that case, we could compute the measure of C given a genotype g and an environment k (x_{gk}) as

$$x_{gk} = \mu + \gamma_g + \varepsilon_k.$$

That is, the particular value of C as determined by environment e_s and genotype g_t is the mean value of C in the population (μ) taken together with the effects of the particular genotype and the particular environment. Things do not work out this way in reality. In fact, there is no single effect a genotype will have on C and no one effect an environment will have on C. The genotypes and environments interact with each other nonadditively. In different environments, different genotypes result in different effects on C. Therefore, in a wooded environment, genotype 1 results in a greater value to C than genotype 2. In a prairie environment, genotype 1 results in a lesser value to C than genotype 2. In that situation, there are no constants one can determine that express a linear relationship between genotypes, environments, and values of C. This is the situation we usually find that we face. Empirically, it seems that nonadditivity is not rare but rather common. This means that there is, strictly speaking, no definable notion of heritability in-the-broad-sense for C in such commonly arising circumstances and no answer the question on how heritable in the broad sense is C in the population we are considering.

GENETIC HERITABILITY IN THE NARROW SENSE

Heritability in the narrow sense is a notion developed that aims to determine a measure of the genetic variance in a character but also the contributions of each individual gene in that variation. The situation with heritability in the narrow sense is even more stark. Once again, the problem is that the interactions between different genes and between different genes and the environment is not additive. For that reason, it is simply not possible to define a measure of the contributions of the different genes. What is sometimes done is to suppose that the interaction between the genes and between the genes and the environment is either a constant (and so is additive) or the special case of zero (for ease of calculation). The world does not provide real-life cases that fit this happy assumption. Therefore, in most cases, there is no measure of heritability in the narrow sense that is applicable. This is not just a problem in finding out what that measure is. It is not that we find it hard to determine what the measure of heritability in the narrow sense is given the evidence we have at our disposal. Rather, it is that there is no such measure; in the same way that a table with nonparallel sides does not have a length. If we assume that the sides are parallel, then we can find a measure for its length but if what the real world has is not like that (if the sides are not parallel), then all we have is a fictive measure.

These two latter notions of heritability can be defined and will give a measure that can be used for certain purposes but only under very strict conditions that, in real biological situations, very often do not apply. It should be clear that just because we do have genotypes and environments interacting to produce a particular value in phenotypic characteristic, C does not mean that it is inevitable that there will be a measure of the genetic causes of resemblance between parent and offspring (heritability in the broad sense). Nor that there is a measure of the effects on the variance of a character of each individual gene (heritability in the narrow sense). These two notions do not play the same theoretical role that the correlational conception of heritability plays.

The simple correlational notion gives a measure of the predictability of offspring having a characteristic given the parents have the characteristic. For that reason, it is a notion that is well defined over a broad range of cases and that can be used to make sense of changes in frequency of individuals with that characteristic. If that characteristic is strongly heritable, in that sense, and if that characteristic increases fitness, then we would expect, other things being equal, that the relative frequency of individuals with that characteristic to increase. This was, in effect, the simple argument Darwin gave for natural selection.

On the correlational understanding of heritability, the heritability of a trait is the probability of an individual having the trait given that the parents have the trait. It is because we can characterize the notion like this that allows us to look at the property of traits that is reflected in their persistence through time in a population. Highly heritable traits are in this sense "passed down" from generation to generation. These traits are strongly correlated between generations. This also makes the heritability of a trait something we can discern by looking at patterns of trait exemplification across time. Do we find that there is a great tendency to find the trait in offspring of trait bearers? If so, then it is heritable. If not, then it is not heritable.

It is not the aim here to argue that the other notions of heritability are not useful in their sphere of relevance. Although as Jacquard (1983) has shown formally and as we have seen previously, the mathematical assumptions made in defining these notions have biological assumptions that are often false. This ensures that heritability in the broad sense and heritability in the narrow sense only apply in very constrained circumstances; circumstances much more constrained than their popularity might suggest. The lesson from this is that the notion of heritability, as it is used in the theory of natural selection, does not mention a mechanism and does not need to mention a mechanism. There need to be mechanisms that undergird the empirically discernible correlations that allow for natural selection but we do not need to know what they are to use the theory of natural selection. What the theory requires is merely the persistence of a characteristic across generations, a characteristic in Darwin's case that increases the degree of adaptation. That notion is the notion of heritability characterized as a correlation between the parents having the characteristic and the offspring having the characteristic.

As Darwin emphasized and after him Lewontin (1978) suggested, what makes a characteristic liable to natural selection that it is adaptive and that it is heritable. It is no part of that story that the characteristic is genetic, even in the attenuated sense; we have defined in terms of its being relatively insulated to environmental changes. Notice that the attenuated sense of "genetic" characterized above required the features to be displayed over a range of environments. It can happen that a population is only found in a very limited range of environments and that those are environments in which an adaptive trait is displayed. It does not follow from that fact that the trait would be displayed across a range of environments, and so it does not follow that the trait is genetic in the attenuated sense.

Feldman (1992: 151) suggests that the notion of heritability was introduced "to quantify the level of predictability of passage of a biologically interesting phenotype from parent to offspring." The idea of heritability as correlation suggested is in keeping with Feldman's, stripped of the idea that the phenotypic trait has to be "biologically interesting." There is no reason to think that only the features that are biologically interesting are heritable. That seems an unnecessary accretion.

GENES AS A MECHANISM, NOT THE MEANING OF HEREDITY

It is an important feature of this correlational notion of heritability that it can be discerned independently of the mechanisms that bring it about. We can find out that a trait is heritable and not know how that heredity is undergirded. That is, we can discover the correlation without knowing why the trait is correlated between parent and offspring. This is where our understanding of genetics does come in: as a mechanism that explains the heredity of traits. Additionally, this was something notably absent from biology at the time Darwin wrote his *Origin of Species*. He could see that many traits that varied in a population were heritable but he did not have a mechanism that explained just how that heritability was possible. The attempt to define heritability in terms of genetics is a change of subject from the phenomenon of heritability itself to a property of a mechanism of heredity.

The heritability of a trait is a measure of the probabilistic relationship between the offspring having the trait and the parents having it. Genetics provides one mechanism that, in the normal run of things, allows us to understand how a parent, which has a certain trait, can have offspring with that trait. It is not the only such mechanism. Discovering that a particular trait is heritable in a population does not settle what mechanism, or mechanisms, explains that persistence of the trait from parents to offspring.

Even though it is itself a probabilistic relationship, we can still raise the question of whether this measure of the heritability of a trait is given absolutely or whether it is relative to population and environment. Just because a trait occurs in a population with a certain heritability, does that show us that the trait will have that heritability in all populations? No, clearly that is not guaranteed. Therefore, in general, heritability will be defined for a population in an environment. Whether we can generalize from heritability in one population to beyond this is an important question about the contingent facts surrounding the persistence of traits across generations and cannot be settled by definition.

When we do our biological research, we are usually dealing with populations in the environments we find them in, and when we say that a trait is heritable to degree n in a population, we mean in the population as we find them, and thus heritability is relative to the environment and the population. It is possible, in principle, that there should be a trait that is heritable to one degree in one population and to another degree in another population.

We have seen that saying that a trait is highly heritable does not entail that it is genetic. We saw how that phrase is multiply confused. However, focusing on the case of mastery of English can also show us that heritability is a property of traits that are far from widely insulated to changes in environment. After all, mastery of English is not insulated to wide changes of environment and yet it is heritable because the normal situation is one in which English-speaking parents raise and speak to their offspring and is such that the probability that an individual has mastery of English is increased by the evidence that that individual was raised in the normal manner by English-speaking parents.

That a phenotype is widespread, or even that a phenotype presents universally in a species, does not mean that it is genetic even in the attenuated sense we allowed earlier. For a phenotype to be genetic in that sense is for it to be relatively insulated to changes in the environment. However, the environmental range for that species might be relatively narrow. Therefore, it simply would not follow that such a widespread phenotype should be thought of as genetic in our attenuated sense.

That a characteristic is heritable, that it is widespread, and even that it is adaptive does not entail nor require that the characteristic is relatively insulated to changes in the environment.

WHAT HERITABILITY TELLS US ABOUT THE GENETIC PERSPECTIVE ON EVOLUTION

Dawkins argues that we are forced to regard genes as the units of selection because they alone meet what he regards as the *sine qua non* criterion of fitness: the targets of selection must be the sort of thing that persists over many generations in which links between the generations persist identically or nearly identically. This is why Dawkins suggests that chromosomes are not candidates for units of selection because they change too frequently through crossing over and genes stand out as prime candidates. We can see this is just not true. What is true is that some phenomena need to be explained at the genetic level, the difference-making characteristics are properties of genes. However, similarly, there may be other phenomena in which genes play a causal role without being the difference-making characteristics. With adaptive phenotypes, there can be what Waddington (1942) called canalization and what is now investigated as robustness of the phenotype (Kitano, 2004; Wagner, 2008; Masel and Trotter, 2010). This process is what allows different genetic variants to cause the same adaptive phenotypes.

WHAT DOES EPIGENETICS DO TO THE THEORY OF NATURAL SELECTION?

Recent developments have suggested, with very firm evidence, that the picture of the way genes play a role in inheritance had been too simple. The neo-Mendelian picture of Darwinism is essentially gene centered. The picture that Mendel had developed was that there was a unit of inheritance called a "factor," which he identified by its effect on phenotype. Therefore, he distinguished between a factor for color and another for shape of a pea flower. Mendelian factors mapped one–one with features. That account did not last long. First, there was the recognition that some features depend on many genes. However, there was also the recognition that genes have consequences for many features. Thus, the relationship between genes and traits is many–many rather than one–one. The picture has been that genes are transcribed into long sequences of amino acids, which then fold into proteins. The process leading from such transcription of genes to complete phenotypes is clearly not a simple one but for a long time the process was regarded as essentially deterministic. Multiple studies seem to suggest that the picture needs to be made more complicated. Either we need to complicate the notion of genotype by including not just the nucleotide pair sequence but also such features as chromatin marks (the mechanisms thought to explain the altered pattern of gene expression). Indeed, we could hold to the notion of genotype that we have but recognize a richer notion of environment. The environment might need to encompass the history of the ancestors of the individual and not just the immediate environment of the individual. We had thought that such history could only affect the individual by bequeathing a genotype. Now, we have grounds to believe that there can be other ways that history can affect the development of that individual.

Previously, we saw that there is no mapping from genotypes to phenotypes because, relative to different environments, a particular genotype produces distinct phenotypes. Epigenetics, or a more indeterministic understanding of the processes of ontogenesis, leads to the thought that the same genotype, even given the same environment, produces distinct phenotypes depending on the factors that shape the way the genes are expressed. These factors will include such things as past episodes of famine, exposure to environmental cues such as chemicals, and so on. Presumably, we will discover more and more of these environmental factors that impinge on the expression of genes. One thing to note is that the idea that one genotype can lead to distinct phenotypes in the same environment requires that notion of environment to be very strictly limited. After all, are we forced to say that if twins had been exposed to distinct environmental cues in the past, and then placed in a similar environment, that they develop in "the same environment"? Not really, this is a matter of stipulation and we are not forced to say that at all. Similarly, it may turn out to be important to include in the notion of an environment not just the chemical environment of the developing fetus but also a more extended environment including that of its predecessors. Thought about that way, one lesson of epigenetics is that we must guard against being too quick to decide what environmental factors are playing a causal role in interacting with the genotype to determine the phenotype. It ought to be clear though that none of this changes the basic idea that heritability will involve the interaction of genotype and environment.

The upshot is that because epigenetics changes only our conception of the mechanism or explanation of the heritability of characteristics, not whether they are heritable, there is no more need to change the theory of natural selection than Darwin did when we accepted, as he did, cases of inheritance of acquired characteristics. We seem to have more solid evidence for the epigenetic effects than Darwin had for the inheritance of acquired characteristics, but neither would fundamentally change the sort of explanations Darwin's theory of natural selection offers. In each case, all that matters is that the characteristics are heritable and that they confer adaptive advantage.

THE NATURE/NURTURE DEBATE

What do we make of the nature/nurture debate accepting this perspective on heritability? One thing that stands out is what it is that we are interested in when we are concerned about the nurture/nature debate Kempthorne (1957, 1978). It is about the ways in which characteristics are insulated to various interventions, whether that is genetic or environmental. Is it true that thugs will be thugs, that geniuses will be geniuses, or does the development of these characteristics depend on the coming together of particular circumstances we can avoid in one case and foster in the other? Is it true that the disease will develop regardless of what the patient does or not?

These questions are about the insulation of the characteristic to changes we can make to the particular genotype and environment pair they have been dealt. We have known how to change the environment and also that such changes can sometimes make the difference we want for millennia. We are now realizing that we can make changes to the genotype too. It is also true that when we are dealing with a particular individual, the easiest way of making a change is a change to their environment, in the broadest sense. Changes to their genotype are not as tractable a solution.

By focusing on the counterfactuals that capture the ways in which phenotypes are insulated to variations in genotype and environment, we get a sense of which manipulations we can influence (by varying the genotype and environment) that can have an effect on the phenotype. This does not mean that any feature is entirely the effect of environment, nor is any feature entirely the effect of genotype. After all, for all phenotypes, there will be a range of genotype–environment pairs that lead to it. In practice, we find ourselves with particular individuals with particular phenotypes. When we want to effect a change in these phenotypes, only some manipulations are available to us. Partly, this is about the limitations in our knowledge and resources. Partly, this is about the cost and benefit of such changes. Inducing a change in one direction may have undesirable effects as well as the desirable effects we seek.

ENGLISH IS BIOLOGICALLY HERITABLE

Let us return to the contested and slightly outrageous thesis of this chapter: English, or the mastery of it, I claimed is strictly biologically heritable. Therefore, we now know what that means and we know what has to be the case when it is true: is it true that in the normal situation, there is a strong correlation between a parent having that characteristic and the offspring having that characteristic? Well, I have not done the study but it would surprise me no end to find out that it is not the case. However, this is not unique to English. It is a commonplace truth that the child of Icelandic-speaking parents, in the normal situation in which they are raised by such parents, will have a great probability of having mastery of Icelandic. Hence, it is no mystery that Icelandic is spoken mostly by those raised by Icelandic speakers. Of course, this is the normal case and is overwhelmingly expected. There are deviations from the norm and in those cases we look and expect to find explanations for intervening factors that stop the child from mastering the language. Hearing problems and underlying brain structural problems, for example, can play these intervening roles. This we would expect to apply to Cantonese, Swahili, and Amharic, etc. It seems plausible for Icelandic, for Cantonese, and for Swahili that the probability of the offspring having mastery over those languages is high given that they are raised by parents having mastery over those languages. In that case, those languages are strictly biologically heritable. The same is true for English.

Once we see that a mastery of English is biologically heritable, while not being "in the genes" either in the sense of being a genetic trait or in the attenuated sense, we have characterized previously in terms of relative insulation to changes in environment, a number of things follow.

As far as research goes, the interest should be focused on what sorts of variation in genotype and environment the development of the phenotype in question, mastery of English, is insulated against. We know that over a wide range of human genotypes, this trait is able to be manifested. On the other hand, we also know that there are a wide range of environments over which the phenotype is developed but that these share the social practice of use of the English language. Therefore, the heritability of English is underpinned by the prevailing practice of using that language.

The claim that mastery of English is heritable does not make any claim about the genetic character of heritability and thus it does not say that it is heritable in the broad sense or in the narrow sense. The claim that mastery of English is heritable is exactly the claim that it is heritable in the correlational sense. The crucial thing is that this is the important sense for natural selection. If the characteristic makes a contribution to the fitness of its bearer, then what matters is that it is present in the offspring howsoever that is achieved, and that is what the correlational sense of heritability indicates.

If a characteristic were heritable in the correlational sense and if it were adaptive in that it makes the parents fitter, then, being heritable, it is going to be present in the offspring and it will, presumably, play a role in making them fitter too. Were mastery of English a characteristic that conferred an adaptive advantage, it would be liable to natural selection. That is, individuals with that trait would be liable to survive differentially because they had that trait.

The mere fact that a trait of a biological organism is heritable does not mean that it is a genetic trait in the strict sense of being a property of the individual's genotype, or that the trait is genetic in the attenuated sense of being a trait that is relatively insulated to changes in the environment. Neither does it follow from a trait's being heritable in the correlational sense that it is heritable in either the broad or narrow sense used in population genetics. It may be heritable in one or both of those ways but that is not required for the trait to be heritable in the correlational sense. Moreover, as we have seen, that is not where the explanatory power for the theory of natural selection resides. Mastery of English is a heritable trait that depends on exposure to English speakers, after all.

Finally, because a characteristic is liable to being selected if it is adaptive and only heritable in the correlational sense, the fact that a phenotype is widely present or even universal in a population is no reason to think that it is genetic even in the attenuated sense spelled out earlier. One of the really important questions which that question turns on is one that biologists cannot answer because it relates to the nonexemplified <genotype, environment> pairs. To see if a characteristic that is universal in a population is genetic in the attenuated sense, we need to see whether other <genotype, environment> pairs deliver the characteristic in question, but we do not have those other pairs exemplified. On the other hand, questions of the adaptive significance of the trait can still be pertinent regardless of whether the trait is genetic in either sense or indeed whether the trait is heritable in the correlational sense we have been discussing. However, if we seek to explain the presence of the trait by using the theory of natural selection, it is important that we can establish both the adaptive significance of the trait and that it is heritable because natural selection depends on the heritability of traits with differing adaptive significance. However, although the heritability of traits is one thing, the mechanisms underlying that heritability are quite another.

REFERENCES

Anderson, S. R. and D. W. Lightfoot. 1999. The human language faculty as an organ. *Annual Review of Physiology* 62: 697–722.

Bailey, D. B. Jr., D. Skinner, and S. F. Warren. 2005. Newborn screening for developmental disabilities. *American Journal of Public Health* 95: 1889–1892.

Dunbar, R. 2000. On the origin of the human mind. In *Evolution and the Human Mind,* edited by P. Carruthers and A. Chamberlain. Cambridge: Cambridge University Press. pp. 238–53.

Feldman, M. W. 1992. Heritability: Some theoretical ambiguities. In *Keywords in Evolutionary Biology*, edited by E. Lloyd and E. F. Keller. Cambridge: Harvard Univ. Press. pp. 151–157.

Guralnick, M. 1998. Effectiveness of early intervention for vulnerable children: A developmental perspective. *American Journal of Mental Retardation* 102: 319–345.

Hurford, J. R. 1999. The evolution of language and languages. In *The Evolution of Culture*, edited by R. Dunbar, C. Knight, and C. Power. Edinburgh: Edinburgh University Press. pp. 173–194.

Jacquard, A. 1983. Heritability: One word, three concepts. *Biometrics* 39: 465–477.

Kempthorne, O. 1957. *An Introduction to Genetic Statistics*. New York: Wiley.

Kempthorne, O. 1978. Logical, epistemological and statistical aspects of nature–nurture data interpretation. *Biometrics* 34: 1–23.

Kitano, H. 2004. Biological robustness. *Nature Reviews Genetics* 5(11): 826–837.

Knight, C., M. Studdert-Kennedy, and J. R. Hurford. 2000. Language: A Darwinian adaptation? In *The Evolutionary Emergence of Language: Social Function and the Origins of Linguistic Form*, edited by C. Knight, M. Studdert-Kennedy, and J. R. Hurford. Cambridge: Cambridge University Press. pp. 1–15.

Lewontin, R. C. 1978. Adaptation. *Scientific American* 239: 156–169.

Masel, J. and M. V. Trotter. 2010. Robustness and evolvability. *Trends in Genetics* 26(9): 406–414.

Schweitzer-Krantz, S. and P. Burgard. 2000. Survey of national guidelines for the treatment of phenylketonuria. *European Journal of Pediatrics* 159(Suppl. 2): S70–S73.

Seashore, M. R., R. Wappner, S. Cho, and F. de la Cruz. 1999. Management of Phenylketonuria for Optimal Outcome: A Review of Guidelines for Phenylketonuria Management and a Report of Surveys of Parents, Patients, and Clinic Directors. *Pediatrics* 104(6): e67.

Waddington, C. H. 1942. Canalization of development and the inheritance of acquired characters. *Nature* 150(3811): 563–565.

Wagner, A. 2008. Robustness and evolvability: A paradox resolved. *Proceedings of the Royal Society B: Biological Sciences* 275(1630): 91–100.

8 Concluding Remarks

Understandably, genetic reductionists scratch their heads and wonder why people are concerned by their remarks about the gene's point of view on evolution. G.C. Williams, and the many others who followed him in taking this perspective, did not think that genes determined phenotypes on their own. They understood as well as anybody the mantra biologists are all taught, that the phenotype is the outcome of the interaction of genotype and environment. Yet people still felt uncomfortable with the pronouncements of the genetic reductionists, and so they should. We have seen reasons to think that the position that has been called genetic reductionism is unsustainable unless we are genetic determinists. The biggest problem is that since genetic reductionism is an attempt at an explanation of the dynamics of the biota and any explanation that only examines things at the genetic level, even if it takes environment into account, it will miss the phenomena that shape the changes the biota goes through if they do not counterfactually depend on features of the genetic level. This is not an *a priori* argument. It depends on the many cases we have seen where the effects of genotypic variation is nonadditive so that we cannot define either heritability in the broad sense or heritability in the narrow sense. The possibility that is *a priori* is that this is possible: a cause of a given event need not be what that event counterfactually depends on. This is a general point about the relationship between causation and counterfactual dependence. It is not specifically about the genes that are part of the actual causal network that brings about the phenotypic characters and so the different tendencies to survive of their bearers.

The gene's point of view can chart the changes in the biota, but that does not mean it can explain them. We can chart changes in a game of baseball by charting the positions of players. This will not explain why the changes take place. Yet any change in the game will involve players being in different locations. When we are seeking to understand why the game evolved the way it did, we need to know the features that caused the changes; player positions are often caused by changes rather than causing changes. A dropped catch will cause different changes than a good catch.

Another analogy is actually discussed by Plato (*The Republic*, 514a–520a) in his allegory of the cave. In this allegory, Plato supposes that there are prisoners from childhood shackled in a cavern, who face a wall on which shadows play cast by people and puppets. These prisoners are no more stupid than you or me; they find patterns in the play of these shadows. They can develop theories about the way one type of shadow brings others or keeps others away. They develop explanations for what they see; they explain the changes they observe in terms of the shadows they see. In doing this, they are mistaken. Shadows do not cause other shadows to fall or run away. It is what casts the shadow that causes something to fall, and so its shadow falls too. The prisoners get the theory of the world wrong. They do not after all explain what they seem to explain. It is not that the prisoners are misled by an illusion. They are seeing a part of reality; the shadows are after all real. What is going on is that they mistake what they see as explaining what they see. The shadows do chart the changes in that small part of the world. The changes in the shadows do not however chart the causes of the changes in the shadows. Changes in shadows may reflect the causes, there can be no change in the shadows without a change in the causes of the shadows, but the changes in shadows are not themselves the causes of the shadows.

This is exactly what happens in evolution with respect to gene frequencies. Looking at changes in gene frequencies will show the effects of evolution. The question (and its answer is not *a priori*) is whether the changes in gene frequencies explain the evolutionary event or not. There are just lots of different cases here. There are some cases where the explanation is at the genetic level and not reflected at other levels. These are cases such as meiotic drive. In this phenomenon, some varieties of gametes from heterozygote parents are either over- or underrepresented in gametes. This can lead to an overrepresentation of less adapted individuals in the population. This is clearly a phenomenon at the level of gametogenesis, and the object that this mechanism gives a tendency to persist is the genotype that is overrepresented. Some evolutionary phenomena can only be explained at the genetic level. This is in keeping with the argument given here. What is being selected depends on what increases in the tendency to persist, that is, what the survival counterfactually depends on. Sometimes this is the gene, and sometimes this is a phenotypic feature of an individual. It is even conceivable that this can be a feature of an object that is bigger than an individual, a group, or a hive. That depends on whether the tendency of that object to persist counterfactually depends on the feature in question and that it is a heritable feature in the correlational sense we explored. At the same time, there can be selective pressures leading in different directions at different levels.

Darwin saw his theory as providing a solution to the "species problem." He saw that there was a causal mechanism hiding in plain sight. The premises he used were known to all who had grappled with the species problem. This is not merely a charting of changes; it is explaining how the changes can take place. The explanation makes use of the plain fact that organisms are well adapted to their manner of living to explain that very fact. If any variety arises that is better adapted, then that variety will tend to survive more frequently, and if the differences are heritable, it will tend to leave more offspring. So it goes. That is, the familiar argument we saw is convincing and powerful. So why is evolutionary theory so hard and so easy?

Doing evolutionary science is easy because we start with organisms that are manifestly adapted and considered at the gross level of comparison: Tuna are adapted to their pelagic lifestyle and would seem to be unsuited to a world without open oceans. We can often specify how features relate to the environment in which the organism is found. For this reason, we often find ourselves reverse engineering from the features to the problems those features solve. This is where the difficulty comes in. We cannot directly measure the tendency to survive of any type of organism. We can make the simplifying assumption that what we see in terms of actual survival numbers accurately reflects those tendencies to survive, but we know that this simplifying assumption is false. We know that the world is more complex than that. As we have already seen, a biblical writer said, "The race not to the swift nor the battle to the strong." These tendencies are not always manifested in direct proportion to their strength. Even more problematic is that we have no direct way of assessing the tendencies, so we cannot get a clear idea, in general, what these tendencies are and how much the actual survival numbers are deviating from a pattern that would exactly match the tendencies. If we ignore this complication, no one could tell. That is the easy bit. The hard bit comes in when we want to know whether there is a difference between survival patterns that we observe and those that march in step with the tendencies. This problem is the problem of identifying the fittest. There is another problem that comes in even when we have identified the fittest in justifying the claim that they survived because of the features that made them the fittest. This is a causal historical thesis, and that makes it particularly hard to justify. In fact, we often work backward here. We see a variety has survived, so we surmise it was fitter and then try to figure out why it was fitter and how that would have explained its differential survival. This is why so many cases of natural selection in the wild presuppose they are looking at cases of natural selection rather than showing that they are cases of natural selection.

We could be justified in using the theory in such a wholesale manner were adaptationism to be true. As a general claim, that all features are optimal or part of the optimal overall solution, adaptationism is false. However, the explanations provided with that assumption need not be false, or at least need not be false in every instance.

The problem with adaptationist explanations is not that they are false. We often have no idea if they are true or false. They may be often false, but they could sometimes be true. The real problem with adaptationist explanations is that they are like astrology. Astrology makes predictions on the basis of evidence that has nothing to do with the predictions. It would be nice if we could assure people that the predictions the astrologer offers are false, but we cannot. What we can do is assure them that the predictions are made on the basis of no compelling evidence at all. So it is with adaptationist explanations. Adaptationist explanations float free of the sorts of grounded compelling evidence that forces us to accept them. For that reason, they are often poor science. They reflect that adaptationism is a presupposition of research rather than the result of research.

The pattern goes like this: There is this characteristic in a population. It must be adaptive. So what is it about that characteristic that is adaptive? And if it is adaptive, it must have arisen by natural selection for that characteristic, and there must be a gene for that characteristic. Each of those steps is precarious and liable to go wrong, and yet the way evolutionary theory gets used makes the steps seem obligatory. Not only are they not obligatory but in fact, we obscure the way the theory works by thinking that way.

We do not need to be adaptationists to use Darwin's theory. We have seen that the general form of the theory is far more robust and powerful than those who take the gene centered view of evolution notice. The notion of heritability required for Darwinian explanations is far more general than the genetic story we have learned. The genetic story we have discovered, and it is a remarkable discovery, is the discovery of a very powerful mechanism of heredity. That story has explained one of the fundamental premises Darwin used in his key argument for natural selection: the heritability of variation. It does not justify that premise. That premise did not need further justification. The heritability of variation was already empirically established in Darwin's day. Darwin was inevitably cautious in stating his other premise. If some variety were better adapted, then it would tend to be more successful in surviving to reproduce, he tended to say. It seems to me that he used that conditional form for a reason. Being better adapted is not easy to discern empirically. We tend to use what we can measure, such as reproductive success, as evidence of fitness. The short step from taking what we can measure as evidence to thinking of it as defining the concept is a massive mistake. For we fall into the circularity objection in that case. There seems to be a pattern here. There is disquiet about the lack of operational or empirical definition for Darwin's key terms. Fitness or degree of adaptation can be understood well enough but turn out not to be measureable. What we can measure, reproductive success, gives us evidence of fitness, but it is a defeasible evidence. Hence, the empirically minded scientist might rebel from Darwin's conceptual appa-ratus and insist on replacing his notion of fitness with something with clear empirical application conditions. The concepts available to do that gut the theory of its explan-atory power, as we have seen. The justification for the theory is its explanations and the way it follows from Darwin's key argument. The gutted theory has neither of these properties.

Where do I stand? I stand with the difficult to apply but explanatory theory. We should expect that these difficulties will not disappear. Our attempts to make sense of the present by charting what happened in the past and importantly to say why it happened is never going to be easy. We know we can get these accounts wrong. Still it has to be said. The theory Darwin presented is simple, powerful, and beautiful. It allows us to understand the world in a way that is consistent with our best contem-porary science and has done so for more than 150 years. Newton did not turn out to be so lucky.

Glossary

Allele: An allele is one of numerous varying forms of a gene.

Canalization: The process leading to the ability of organisms to produce the same phenotype despite variations in genotype and environment. See Waddington.

Catastrophism: The idea that the world has its character largely due to catastrophic events of a worldwide or cosmic scale. This view was prevalent before Hutton's uniformitarianism. The sources of catastrophism are many and varied. The Stoics in ancient Greece argued that the origin and character of the universe was due to cosmic conflagration (ekpyrosis), for example.

Chromosome: Structure that is made of double-stranded DNA. These occur inside the nucleus in eukaryotes and as a circular double-stranded molecule of DNA in prokaryotes.

Darwin, Charles (February 12, 1809–April 19, 1882): English naturalist, Fellow of the Royal Society, developed his theory of natural selection in about 1838, eventually published as *On the Origin of Species* in 1859. Gentleman scientist, never employed but ever working.

Darwinian: Of or pertaining to Charles Darwin; contrast with neo-Darwinian.

Deterministic: Of a world or theory. A world is deterministic if and only if the laws governing that world together with a section of history uniquely determine a complete history for that world. A theory is deterministic if it requires the world to be deterministic for it to be true.

Dispositional property: A tendency, for example, solubility in water is a tendency to dissolve in water; an object can have this property without ever dissolving in water.

Dobzhansky, Theodosius (January 24, 1900–December 18, 1975): Ukrainian foreign member of the Royal Society. Important population geneticist and evolutionary biologist. Worked primarily in the United States.

Epigenetics: Inheritance of characteristics between generations, which occurs without changes of the primary structure of DNA. The term was coined by C.H. Waddington. This takes many forms: between cells or between individuals. This is an ongoing and quickly developing area of research and discovery that complicates the account we have of the mechanisms of heredity.

Evolution: Heritable organic change over time; contrasted with both development and phenotypic changes that are due to environmental changes.

Fitness: In Darwin's use and most often in this book, this is a measure of the tendency of a variety to survive to the point where it can reproduce. Modern biologists often define this property as a measure of reproductive success; cf. inclusive fitness.

Function, theories of

    Adaptive: The biological function of a phenomenon is how it makes the system of which it is a part to have a greater tendency to survive or persist.

Cummins: The biological function of a phenomenon is what role it plays in a system of which it is a part.

Selectionist: The biological function of a phenomenon is the reason why it was selected.

Genotype: The genetic characteristics of a cell or by extension an organism. Although there are extra genetic materials.

Genetic determinism: The doctrine that genotype determines all other characteristics, phenotype.

Genetic drift: Change in frequencies of genetic variants due to random sampling processes. This may occur at the formation of gametes or at various other points.

Genetic reductionism: The doctrine that fitness and the causal explanation for evolution are uniquely properly understood from the genetic level of analysis. Fitness, for example, is directly a property of genes, not individuals and not traits.

Gosse, Edmund (September 21, 1849–May 16, 1928): English poet and critic, son of Philip Henry Gosse. Author of *Father and Son* (1907), an account of his relationship with his parents and in particular his father. Early lecturer on English literature at Trinity College, Cambridge, in the 1880s.

Gosse, Philip Henry (April 6, 1810–August 23, 1888): English self-taught naturalist and expert in particular in aquatic ecology. He invented the seawater aquarium to facilitate his research and sparked the aquarium craze. He was commissioned to build and stock a seawater aquarium for the London Zoo, which he delivered in 1853. This was the first public aquarium. He was also a deeply religious person with literalist convictions. His attempt to reconcile his scientific views with his religious convictions resulted in his book *Omphalos* in 1857. This title comes from the Greek for belly button. He argued God would have created Adam with a belly button, that is with an *as if* history. He also argued the world would have an *as if* history.

Gould, Stephen Jay (September 10, 1941–May 20, 2002): American paleontologist and evolutionary biologist.

Hamilton, William Donald (August 1, 1936–March 7, 2000): English evolutionary biologist. Instrumental in the development of a gene-centric perspective of the evolutionary process.

Heritability:

Correlational: A measure of the correlation between parents having a (phenotypic or genotypic) trait and offspring having it.

Heritability in the narrow sense: Proportion of variance in a character which can be assigned to distinct genetic loci.

Heritability in the broad sense: Proportion of variance in a character due to the variance in genotype.

Hutton, James (June 3, 1726–March 26, 1797): Scottish physician, farmer, amateur investigator and founder of what we now know as geology. He suggested that the explanation for geological phenomena could be found in forces visible today. This was called "uniformitarianism" by William Whewell. This required much more time than literalists of the Bible thought possible. This was a blow to those who sought support on the literal truth of the Bible.

Inclusive fitness: A measure of evolutionary success. The idea, due to W.D. Hamilton (1963, 1964), that the evolutionary success (inclusive fitness) of an instance of an individual gene is to be measured not simply by the number of descendants of that instance but rather by the number of copies of that gene that are caused to exist because of it.

Lamarck, Jean-Baptiste (August 1, 1744–December 18, 1829): French aristocrat scientist who developed an early theory of evolution that emphasized the causal efficacy of final ends. He is mostly remembered today for the doctrine of inheritance of acquired characteristics, a doctrine he held but was widely held at the time.

Lamarckism: A theory of evolution that involves the causal efficacy of the goal or telos of evolution developed by Jean-Baptiste Lamarck. This is the idea that the goal of evolution explains the steps on the way to that goal in a manner in which our own intentions explain the steps we take to achieve them.

Lewontin, Richard Charles (March 29, 1929): American biologist and population geneticist.

Lyell, Charles (November 14, 1797–February 22, 1875): English geologist, Fellow of The Royal Society. Author of *Principles of Geology*, which used James Hutton's ideas of uniformitarianism. In this, he influenced Charles Darwin to give an account of changes to the biological world that fit the uniformitarian principles.

Malthus, Rev. Thomas (February 13, 1766–December 29, 1834): Fellow of The Royal Society. Worked mostly in political economy.

Malthusian: Associated with Reverend Thomas Malthus, who published *An Essay on the Principle of Population* (1798) and argued that population growth when unchecked is exponential whereas food supply can only grow arithmetically. This idea suggests that there will be competition for food and similarly for any limited resource. This idea influenced both Charles Darwin and Alfred Wallace in their development independently of the theory of natural selection.

Meiosis: Process by which a diploid eukaryote cell produces a haploid gamete or sex cell.

Mitosis: Process by which a diploid eukaryotic cell produces two diploid nuclei within the same cell. This usually leads to cytokinesis, or the formation of two distinct cells.

Natural selection: Mechanism bringing about evolution, described by Charles Darwin as "the main but not the exclusive means of modification": the differential survival of the fittest because of the characteristics that make them fitter, helping them to survive.

Neo-Darwinism: The "modern synthesis" of Darwin's theory of evolution by natural selection and Mendel's theory of genetic inheritance.

Phenotype: The sum of an organism's properties except for the narrowly genetic characteristics included in the genotype.

Probability: Measure of the likelihood that an event will occur.

    Probability (long-run frequency): An account of probability that identifies probability with the ratio of positive instances and instances in general; for example, the probability of tossing a head on a coin is the number of heads tossed divided by the number of coin tosses.

Probability (subjectivism): An account of probability that identifies probability with the degree of belief.

Probability (propensity or objective chance): An account of probability as an objective feature of the world; propensities are not observed relative frequencies but rather properties of the underlying system that explain the observed relative frequencies.

Simpson, George Gaylord (June 16, 1902–October 6, 1984): Paleontologist and major theorist of the modern synthesis of Darwinism and genetics, known as "neo-Darwinism."

Survival of the fittest: Subtitle to Chapter 4 of *On the Origin of Species* from the fifth edition onward. The phrase is due to Herbert Spencer but adopted by Darwin as an explanation for natural selection.

Standing property: As opposed to a disposition, a fully manifested property.

Uniformitarianism: An approach to explaining the universe that emphasizes that laws and operations presently acting are uniformly applicable throughout the universe and across time. Contrasted with catastrophism.

Van Valen, Leigh (August 12, 1935–October 16, 2010): American paleontologist and evolutionary theorist. Originated the Red Queen Hypothesis and the Ecological Species Concept.

Waddington, Conrad Hal (November 8, 1905–September 26, 1975): British biologist coined the term "epigenetic" in relation developmental pathways. Another important notion developed by Waddington is "canalization" or the process leading to the ability of organisms to produce the same phenotype despite variations in genotype and environment. This idea is being explored presently in developmental biology under the term "robustness."

Wallace, Alfred Russel (January 8, 1823–November 7, 1913): British naturalist (born in Wales, died in England); variously a professional collector of biological specimens, explorer of new regions, early biogeographer, but most importantly independent discoverer of the theory of natural selection. It was Wallace's letter to Darwin containing his 1858 article "On the Tendency of Varieties to Depart Indefinitely From the Original Type" that sparked the latter into action to publish his theory.

Wright, Sewall (December 16, 1889–March 3, 1988): Seminal American geneticist who was a founder of theoretical population genetics with R.A. Fisher and J.B.S. Haldane. Unlike them, he did not see evolution as occurring at the genetic level primarily, that is, he was not a genetic reductionist.

Index

Species and Systematics

SPECIES: A HISTORY OF THE IDEA
John S. Wilkins

COMPARATIVE BIOGEOGRAPHY: DISCOVERING AND
CLASSIFYING BIO-GEOGRAPHICAL PATTERNS OF A
DYNAMIC EARTH
Lynee R. Parenti and Malte C. Ebach

BEYOND CLADISTICS
Edited by David M. Williams and Sandra Knapp

MOLECULAR PANBIOGEOGRAPHY ON THE TROPICS
Michael Heads

THE EVOLUTION OF PHYLOGENETIC SYSTEMATICS
Edited by Andrew Hamilton

EVOLUTION BY NATURAL SELECTION: CONFIDENCE,
EVIDENCE AND THE GAP
Michaelis Michael
www.crcpress.com/9781498700870

For Product Safety Concerns and Information please contact our EU
representative GPSR@taylorandfrancis.com
Taylor & Francis Verlag GmbH, Kaufingerstraße 24, 80331 München, Germany

www.ingramcontent.com/pod-product-compliance
Ingram Content Group UK Ltd.
Pitfield, Milton Keynes, MK11 3LW, UK
UKHW021122180425
457613UK00005B/186